Y0-CAX-989

THE ANIMAL KINGDOM IN JEWISH THOUGHT

THE ANIMAL KINGDOM IN JEWISH THOUGHT

Shlomo Pesach Toperoff

JASON ARONSON INC.
Northvale, New Jersey
London

This book was set in 11 pt. Berkeley Oldstyle by AeroType, Inc.

10 9 8 7 6 5 4 3 2 1

Library of Congress Cataloging-in-Publication Data

Toperoff, Shlomo Pesach.
The animal kingdom in Jewish thought / by Shlomo Pesach Toperoff.
p. cm.
Includes bibliographical references and index.
ISBN 1-56821-439-1
1. Animals in rabbinical literature. 2. Animals—Religious aspects—Judaism.
I. Title.
BM509.A5T66 1995
296.1—dc20 95-13592

Manufactured in the United States of America. Jason Aronson Inc. offers books and cassettes. For information and catalog write to Jason Aronson Inc., 230 Livingston Street, Northvale, New Jersey 07647.

This book is dedicated to the memory of my dear parents.

מאיר ב״ר משה יוסף ז״ל
שרה חנה בת חיים לייב ז״ל

Contents

Acknowledgments

Here I express my thankfulness to members of the family and friends to whom I owe a debt of gratitude for their readiness to assist me at all times. Without their spontaneous and wholehearted cooperation this book could not have been produced.

To my dear wife, Lily; my children, Joy Toperoff, Naomi Gurwicz, and Meir David Toperoff; and my grandchildren, Rabbi Shimon Gurwicz and Gideon Tzvi Toperoff.

And to my friends Rose Cooper, David Lehmann, Tzvi Stern, Pesach Stiel, Bernard Sunshine, and Mark Wolfson.

Preface

Now that I have concluded writing this book, I must address myself to those who have questioned me about the advisability of writing on the animal world when the human world is beset by so many pressing problems.

It is possible that an incident in my early childhood triggered off my interest in the animal world. I recollect that as a child I saw my mother ע"ה place bread crumbs on the windowsill for the birds to eat.

This exercise was enacted with religious devotion every morning of the week, and my mother would not partake of any food before feeding the birds. It was only later in life that I discovered that my mother acted on the rabbinic dictum that one should feed one's animal before feeding one's self.

This follows the interpretation of the verse in the *Shema* that reads, "and I will give grass in thy field for thy cattle and thou shall eat and be satisfied." This verse clearly states that we should feed the animal and only then "eat and be satisfied."

As I grew older this early incident seemed to fire my imagination, and I became more and more interested in the rabbinic sayings found in Talmud and Midrash.

To whet the appetite of the reader, I deal with several areas of life in which the animal world plays a dominant role, and they are all steeped in Jewish tradition and custom.

We begin with the Jewish child. In ancient Israel, it was customary for the Jewish child to be initiated in the study of Torah with the reading of Leviticus, the third book of the Torah, and not with Genesis, the beginning of the Torah. This custom is based on a fascinating Midrash that declares: "As the animal offered on the Altar must be clean (kosher), so the child whose mind is clean commences to study the book of Leviticus which deals with the sacrificial system" (*Leviticus Rabbah* 7:2).

Thus we see that Torah education for the young is linked with the animal kingdom, and the underlying motive for this custom is associated with the famous maxim "Cleanliness is next to Godliness." Incidentally, the Jewish dietary laws are also based on the same maxim.

The land of Israel, which plays a significant role in Jewish history, is described in the *Tanach* as "a land flowing with milk and honey" (Jeremiah 11:5), both of which are rooted in the animal world.

Honey ushers in the Jewish New Year, and on the eve of Rosh Hashanah we recite a short blessing hoping that we shall enjoy a "good and sweet year." Furthermore, throughout the first weeks of the Jewish year we dip the bread in honey and not salt, which is used consistently throughout the year.

Another product that is supplied by the animal kingdom is leather, which is derived from the hides of animals. Here, too, Jewish custom comes into play.

The blessing "*shehechiyanu*" is recited whenever we eat a new fruit or wear a new garment. However, it is not recited over new shoes made of leather because they required the slaying of an animal (see *Kitzur Shulchan Aruch* 59:13). In addition, the leather straps of the *tefillin* should be produced from the skin of a kosher animal (*Kitzur Shulchan Aruch* 33:3).

Wool is another product derived from an animal. The Bible refers to the shearing of sheep on four different occasions: Genesis 31:19, Genesis 38:12, 1 Samuel 25:4, and 2 Samuel 13:27. Observant Jews enwrap themselves in prayer not with a silk *tallit* but with a woolen one. In this manner, we demonstrate our gratitude to, and dependence upon, the sheep whose wool we use for our religious devotions.

Wool is also mentioned in connection with *shatnes*, as cited in Deuteronomy 22:11: "Thou shalt not wear a garment of diverse kinds, of wool and linen together." The law of *shatnes* is classified as a *chok*, which initially has no valid reason. However, it has been suggested by Henry Knobil that "the basic and primary aim which safeguards the divinely ordained state of nature is that any substance which nature intended to keep apart should not be brought together." He explains that "wool fiber is composed of animal tissues and can be seen as a chemical representation of the animal kingdom while linen is composed, in the main, of cellulose carbohydrate and is the very embodiment of the vegetable kingdom. Vegetable and animal fibers must not be mixed together and used in the same garment."[1]

Peace, for which we yearn and crave, is mentioned in the Book of Isaiah where we read that "They shall beat their swords into plowshares and their spears into pruning hooks" (Isaiah 2:4). This is a majestic and glorious vision, yet in chapter 11, the prophet feels compelled to draw upon the animal kingdom when he prophesies that "the wolf shall dwell with the lamb" (Isaiah 11:6). To understand the full import of this vision we should recall the words of the Rabbis who, in symbolic imagery, portray the nations of the world as ravenous wolves that fiercely attack Israel, the defenseless and innocent lamb. In this vision, Isaiah foretells that in the Messianic Era all the nations will enjoy coexistence with Israel, and peace will reign supreme throughout the world.

Yet, another aspect of peace needs clarification. The Jewish form of greeting is "*shalom*," peace, and the rabbis suggest that *shalom* is the concluding word of the priestly benediction. Therefore, when we greet our fellowman with *shalom*, we present him with an abbreviated blessing.

In this context, we should remember that the Torah equates *shalom* regarding the human race with *shalom* concerning the animal kingdom. In Genesis 37:14 we read that Jacob entrusted Joseph with the mission to inquire after "the welfare (*shalom*) of your brethren and the welfare (*shalom*) of the flock."

We know that the Torah uses words very sparingly. We must therefore question why the word *shalom* is repeated twice in the same verse. I humbly suggest that the Patriarch Jacob was as much involved with his flock as with his sons; indeed, the flock is an integral part of his family.

The Etz Joseph, in his commentary on *Genesis Rabbah* 16:13, states that one must inquire after the well-being of anything from which he benefits. In the final analysis, we would do well to emulate Jacob and inquire after the welfare of our flock. We do not realize the magnitude of our indebtedness to the animal world; we owe much to them.

We consume an incalculable amount of meat, poultry, fish, and eggs every day; we drink millions of gallons of milk, we sweeten our palate with honey derived from the bee, we wear innumerable shoes and boots made of leather that is produced from the hides of animals, and sheep supply us with ample reserves of wool.

We echo the sentiments of the prophet who wrote, "Both man and beast, thou savest O Lord" (Psalm 36:7).

In conclusion, throughout my contacts with people I discovered that a goodly number of boys and girls of school age are intensely interested in the animal world; it is my wish therefore that this book will bring both young and old closer to a better understanding of the animal world through the medium of Torah, which embraces rabbinic and chasidic lore.

Sabbath and Festival

Sabbath

The Sabbath, the weekly day of rest, applies not only to the human beings but also to the beasts of the field, as it is clearly enunciated in the Commandments (Exodus 8:10, Deuteronomy 5:14). In Exodus 3:12 we read: "So that your ox and ass may rest."

The saving of life, which abrogates the Sabbath, applies both to human beings and to the animal world as we see from the following passage in the Talmud, which discusses how far the laws of the Sabbath may be violated for the purpose of rescuing an animal that was in danger. If an animal has fallen into a pool of water, we may bring pillows and cushions to place beneath it, and if it thereby ascends to terra firma, well and good. Against this statement of the law, the following decision is quoted: if an animal has fallen in water, we provide it with fodder so that it should not die. Hence it would seem that we provide the animal with fodder but not with pillows and cushions! There is no contradiction, as the latter refers to a case where it is possible to keep the animal alive with fodder and the former where it is not possible. If we can keep the animal alive by feeding it, we should do so; otherwise, we must bring pillows and cushions. But by doing so, we infringe upon a regulation that renders it unusable! That regulation is an institution of the Rabbis, whereas preventing pain to an animal is a command of the Torah, and what is instituted by the Torah sets aside what is instituted by the rabbis (*Shabbat* 128).

The *Halachah* says:

> He who meets two on the road, one lying under his burden and the other which has thrown off his burden, but none had been found to help him load it again, it is required to unload first in order to relieve the suffering of a living creature and later to help load. This applies only where both instances concern enemies or friends, but if one of them was an enemy and the other a friend, it is commanded to help the enemy first, in order to subdue one's inclination. (Maimonides, *Mishneh Torah*, Laws Concerning a Murderer 13.13)

Below are some laws pertaining to the care of animals on Sabbath and Yom Tov:

On Sabbath one may put food in front of animals or birds that are dependent on one for food. Not only the owner, but anyone can feed them.

It is correct to shake crumbs out of a tablecloth in a place where there is an *eiruv* or where no *eiruv* is needed, even though the birds will eat them.

Dogs are in a special category, and one may put food in front of them, both on the Sabbath and on Yom Tov, even if they are not dependent upon man for their food. One is also permitted on Sabbath and Yom Tov to put food in front of any animal that is hungry and cannot find anything to eat.

One may put food into the mouth of animals that have difficulty taking food for themselves.

Regarding geese, one should, if possible, feed them before Sabbath or Yom Tov to avoid their suffering.

It is forbidden to change the water in an aquarium containing fish on Sabbath and Yom Tov. However, if it is necessary to add water every day, one may do so, even if the aquarium contains aquatic plants, because the fish are dependent upon man for their food.

A fish that has jumped out of an aquarium on Sabbath or Yom Tov may be put back to avoid its suffering.

One is allowed to feed birds in a cage because they are dependent upon one for their food.

One may move a cage from the sun to another place so that the birds may be more comfortable.

On Sabbath, one is permitted to drive a domestic animal such as a cow or a sheep into its pen, and one may then close the gate.

One may have an animal milked by a non–Jew in order to save it suffering.

A farm that has a milking machine may use it on Sabbath or Yom Tov, but the machine must be controlled by a timer.

One may feed medicine to an animal, administer an intramuscular injection, and put ointment on an animal's wound.

Food offered to the beast on the Sabbath should not be weighed or measured.

It is permitted to cut hard cucumber for animals on the Sabbath.

It is permitted to let one's beast stand upon herbage that is connected to the soil, and if it has nothing to drink, it is permitted to tell a non–Jew to bring it water from a well in a semiprivate domain.

Cattle, beasts, and poultry that are raised on one's premises may be fed on the Sabbath.

Festivals

Rosh Hashanah

On the Jewish New Year it is customary to partake of the head of a sheep in commemoration of the ram of Isaac.

Shofar: The ram's horn plays a dominant role in the morning service of both days of Rosh Hashanah. God urges Israel to use the ram's horn that we may remember Isaac in his readiness to be sacrificed unto God (*Rosh Hashanah* 16a).

The psychological impact that the *shofar* had on the minds and hearts of the people is assessed by the Prophet Amos, who cried out, "Can the *shofar* be sounded and the people will not tremble?" (Amos 3:6).

Tashlich: On the first day of Rosh Hashanah a little before sunset, observant Jews congregate at the edge of a collection of running water and recite the last verse of the book of Michah. The rabbis suggested that it is preferable that the water should contain fish, which serve the purpose of reminding us that we are like the fish caught in the net—weak, helpless, and subject to the many ills that beset us in life.

It was the custom to take loaves of bread for the New Year with figures of birds on the top recalling the verse "As birds hover, so will the Lord of Hosts protect Jerusalem."

Yom Kippur

Kaporot: The Expiatory Service was enacted in the home on the morning before the Eve of Atonement Day when the male takes a live rooster and the female takes a live hen and while turning it around over their heads, recite the appropriate words recorded in the *machzor*.

At the present time, when Jews live in apartments and do not keep livestock, we use money that is later distributed to the poor.

Sukkot

Many sacrifices were offered during the seven days, seventy in number corresponding to the seventy nations of which mankind is said to consist and on whose behalf Israel sought atonement from God. In this manner Judaism underscores the spirit of universalism.

On leaving the *sukkah*, we recite a short blessing in which we mention the skin of the leviathan, a legendary animal (see the essay on the leviathan).

Passover

As we know, *matzot* are the salient feature of the Passover festival, which is called *Chag HaMatzot* (The Festival of Unleavened Bread). In its manufacture all kinds of artistic designs were perforated into the *matzot*, such as birds, animals, and fish. Opposition was voiced by rabbis to fancy perforations on the grounds that the women spent too much time in efforts to make their creations very artistic. In the *Book of Delight* by Israel Abrahams, the author writes of seeing *matzot* baked in London in the forms of birds, fish, sheep, and monkeys. He could not recollect whether they were primarily for the Passover festival.[1]

According to the Passover ritual, which is found in the book of Exodus, every Israelite was to set aside a lamb on the tenth of *Nisan*, which was to be kept until the fourteenth and then slaughtered. Today, the Pascal sacrifice is symbolized by the roasted shankbone, which is an indispensable item on the Seder table. It is unique in the sense that other foodstuffs on the Seder table are partaken of; however, the shankbone with its small amount of roasted meat is untouched.

Shavuot

Shavuot is the holiday marking the end of the spring harvest.

The *Omer*, which is a sheaf of the firstfruits of the harvest, and the two loaves share much in common. They are both called a "new meal offering"; however, they are different from each other in that the *Omer* is brought from barley, whereas the loaves are of wheat. Wheat is for human consumption, whereas barley is an animal food (compare *Sotah* 2:1).

Animal Welfare

In ancient Egypt, Rome, and Greece, a limited number of animals were worshiped and even deified, but the vast majority were ignored and entirely neglected. Judaism, however, being a purely monotheistic religion, worshiped neither human being nor beasts; every species of animal life was our concern and according to Jewish tradition we were warned not to hurt even a fly. Animal welfare is a recurrent theme in every age, in Bible, Talmud, and post-talmudic literature.

Bible

As early as Genesis 6:16 we read that Noah was asked to construct an ark with lower, second, and third stories. The middle section was reserved especially for animals, and Noah stringently observed the different feeding habits of the animals under his charge. Indeed he was so preoccupied with their needs and requirements that for a period of time he was denied all sleep (*Midrash Rabbah*). Being concerned with animal welfare can even lead to marriage as we see in Genesis, chapter 24. When Rebecca offers to quench the thirst of the camels, Eliezer, the servant of Abraham, knows that she is destined to become Isaac's wife.

The great institution of Sabbath, the weekly day of rest, is reserved not only for human beings but also for animals as we read in Exodus

20:9, Deuteronomy 5:14. See also Exodus 23:12: "That thy ox and ass may rest."

The following laws of the Torah are self-explanatory and are all rooted in animal welfare. These verses exemplify the biblical requirements of protecting animals:

"And he said unto him: My Lord knoweth that the children are tender, and that the flocks and herds with me give suck, and if they overdrive them for a single day, all the flocks will die" (Genesis 33:13).

A meaningful comment on the above verse is given by Samson Raphael Hirsch, who writes, "The animals which I have with me are required neither for work nor for slaughter, but for breeding, to increase my property. As such, I have to look after them and treat them with consideration for their prosperous development."[1]

"And if a man shall open a pit, or if a man shall dig a pit, and not cover it, and an ox or an ass fall therein, the owner of the pit shall make it good; he shall give money unto the owner of them, and the dead beast shall be his" (Exodus 21:33–34).

"If thou meet thine enemy's ox or his ass going astray, thou shalt surely bring it back to him again" (Exodus 23:4). This verse differs from Deuteronomy 22:1 in that it refers to an enemy's ox or ass.

Whether it be cow or ewe, "Ye shall not kill it and its young on the same day" (Leviticus 22:28).

Samson Raphael Hirsch, in his commentary, makes this remark: "The above verse points to the fact that this trace of humanity renders the animal fit for such representation, and the fact that it has to be taken into consideration shows the Jewish conception of sacrifices in their having the sole purpose of the furthering of human morality, and sharply differentiates it from the pagan idea which sees in sacrifices a killing which gives satisfaction to their gods."[2]

J. H. Hertz writes: "It is prohibited to kill an animal with its young on the same day in order that people should be restrained and prevented from killing the two together in such a manner that the young is slain in the sight of the mother."[3]

"Thou shalt not see thy brother's ass or his ox fallen down by the way and hide thyself from them: thou shalt surely help him to lift them up again" (Deuteronomy 22:4).

Commenting upon the above verse J. H. Hertz writes: "A chassidic rabbi lamented that this law was little observed and he devoted himself to its fulfillment. He was continually to be seen in the streets, helping one man to load his wagon, another to drag his cart out of the mire."[4]

"If a bird's nest chance to be before thee in the way in any tree, or on the ground, whether they be young ones, or eggs, and the dam sitting upon the young, or upon the eggs, thou shalt not take the dam with the young. The young thou mayest take" (Deuteronomy 22:6).

Maimonides suggests that grief should not be caused to animals or birds, and consequently how much more careful we must be that one should not cause grief to our fellow man.

"Thou shalt not plow with an ox and an ass together [when he treadeth the corn]" (Deuteronomy 22:10).

In his commentary on this verse, J. H. Hertz writes, "[An ox and an ass] differ greatly in their nature, in size and in strength; it is therefore, cruel to the weaker animal to yoke them together."[5]

Prophets

In a moving picture, Nathan the prophet describes how the poor man "had nothing save one little ewe lamb, which he had bought and reared; and it grew up together with him and his children; it did eat of his own morsel, and drink of his own cup, and lay in his bosom, and was unto him as a daughter" (2 Samuel 12:3).

Joel the prophet exhibits mercy toward the dumb creatures by saying, "The beasts of the field cry also to you, for the rivers of water are dried up and fire has devoured the pastures of the wilderness" (Joel 1:20).

In the Book of Jonah, we see God's concern not only for the people of Nineveh but also for their cattle (see Jonah 4:11).

The Talmud

The Hebrew term *tzar baalei chaim* (compassion for living creatures), coined by the rabbis, teaches us to avoid the infliction of any unnecessary pain to the animal. Whereas we are governed by the 613 precepts of the Torah, the world at large is ruled by the seven Noachide Laws, one of which is the prohibition not to tear a limb from a living animal. The above law may have been in response to the Bacchanalian feasts, when in orgies of drunkenness and licentiousness, revelers tore limbs from a living animal. However, *tzar baalei chaim* has a wider connotation, for it condemned not only the gladiatorial fights in the Roman arena when slaves were pitted against ferocious animals and torn to pieces amidst the uproarious laughter of thousands of spectators but also all blood sports such as bullfighting, cockfighting, and hunting with hounds, which enjoy the patronage of royalty and aristocracy in Great Britain today. It should be noted that the rabbinic comment on Psalm 1, "Happy is the man . . ." prohibits hunting.

We also remember that Josephus testifies to the fact that "Herod also got together a great quantity of wild beasts and of lions in very great abundance and such other beasts as were either of uncommon strength or of such a sort as were rarely seen" (*Antiquity* 15, 8:1). These beasts were trained either to fight one with the other, or with men who were condemned to death. Truly, foreigners were greatly surprised and delighted at the vast expenses of the games and the great danger of the spectacles, but to the Jews, they were a palpable breaking up of those customs for which they had so great a veneration.

Many of the laws of *shechitah* (Jewish ritual slaughter) are based on the assumption that it is absolutely vital to reduce to a minimum the pain inflicted on an animal when taking its life. With this end in view, the rabbis have enumerated five causes of disqualification of *shechitah*:

Shehiyah: "Delay"—there must be a continuous forward and backward motion of the knife without any interruption.

Darasah: "Pressure"—the cut must be made gently without exercise of any force.

Chaladah: "Digging"—the knife must not be inserted into the flesh but rather must be drawn across the throat.

Hagramah: The cut must be made in a slanting direction.

Ikkur: "Tearing"—the cut must be made without dislocating the windpipe or gullet.

Any of these acts invalidate the *shechitah*, as they inflict excessive pain on the animal (*Chullin* 9a).

It is forbidden to cause any suffering to any living creature. This is biblical law (*Bava Metzia* 32b, *Deuteronomy Rabbah* 6:1).

An entirely different aspect of animal welfare is reflected in the following Mishnah. "Cases concerning offenses punishable by death are decided by twenty-three judges. A beast that commits or suffers unnatural crime is judged by twenty-three, as it is written, 'Thou shalt slay the woman and the beast' " (Leviticus 20:16). Again it says, "And ye shall slay the beast" (Leviticus 20:15) and "The ox that is to be stoned and its owner also, shall be put to death" (Exodus 21:29).

In a similar manner, as the owner is put to death, so is the ox put to death. The wolf, the lion, the bear, the leopard, the panther, or serpent (that have killed a man), their death is also decided by twenty-three judges. "Rabbi Eliezer says: 'If any one killed them before they were brought to court, he has acquired merit,' but Rabbi Akiva says 'Their death is decided upon by twenty-three judges' " (*Sanhedrin* 1:4).

According to the view of Rabbi Akiva, the animal is not summarily disposed of but is entitled to a full trial by a court of twenty-three judges. Furthermore, the above Mishnah implies that the beast is on the same level as man as far as capital punishment is concerned.

In this context, it is worth noting that the rabbis distinguish between a *tam* and a *muad*. A *tam* is an animal that has done an injury and has not repeated the same for three days; in popular jargon, it is a first offender (see Exodus 21:28–36; *Bava Kamma* 1:4). The *muad* is an animal whose owner is forewarned on account of three successive injuries. In this case, the owner is liable to pay full indemnity (see Exodus 21:29). This animal is then considered a habitual offender.

The Synagogue and Art

The synagogue was always devoid of all statues, images, and pictures of humankind in conformity with the spirit of the Second Commandment, "Thou shalt not make any graven image or likeness." However, the animal world is represented in many synagogues by the Lion of Judah embroidered on the curtain of the Ark and engraved on the breastplate covering the mantle of the *Sefer Torah*. There is also a prohibition of making any representation of the Heavenly Chariot (Ezekiel 1:10), which is pulled by a lion, but in some passages it is stated that the prohibition only applies to a representation of all faces together. This was accepted as a rule by Solomon Ibn Adret and is the ruling given in the *Shulchan Aruch* (*Yoreh De'ah* 41:4).

It is interesting to note that Rabbi Meir of Rothenburgh prohibited the illuminating of prayer books lest figures of animals and birds divert attention from prayer. It should also be noted that Isaac of Zaruia condemned the frescoes in the synagogue of Meissen. Some of Rabbi Meir's predecessors were of the opinion that the woven covers of the almemar be decorated with patterns of animals because in the Middle Ages, animals were no longer worshiped as idols. Ephraim ben Isaac the Elder, a German Tosafist, generally permitted animal figures in synagogue decorations despite earlier opinions.

The existence of stained glass windows, just coming into fashion in Central Europe, in a twelfth-century synagogue in Cologne, is well attested. In short, the attitude of the medieval Jew to artistic images was one of indifference rather than hostility.

We also found on the seals of *rosh galuta* (head of the Jewish community of Babylon) a picture of a fly (see the chapter on the fly).

Interesting modern examples of synagogue art are the Marc Chagall windows of the synagogue at Hadassah Hospital outside Jerusalem. Chagall was restricted by the fact that the Jewish faith does not allow the representation of the human figure. However, as one who always handled birds, fish, fruit, and flowers with a rare tenderness, he took over the Old Testament images intact, seeing Judah as "a lion's whelp," Dan as "a snake in the way," Naphtali as a "hind let loose," and Issachar as a "strong ass crouching down between two burdens." For each of the big panels (measuring 11 feet by 9½ feet) he chose a different predominant color.

Outside the synagogue, there are some interesting examples of art in the form of animals, such as an ancient Palestinian lamp filler made of pottery in the image of a dog, a spice box in the shape of a fish (Eastern Europe, 19th century), and a container in the form of an ostrich with a horseshoe in its beak, which was made in the seventeenth century in Germany. In the Ukraine in the nineteenth century, we find a box shaped like a hen with hinged wings.

The animal world has even invaded the cemetery; for example, in Prague the headstones in the Jewish cemetery were engraved with animals. The stag was represented in the name Hirsch (Tzvi), a carp for Carpeler, a cock for Hahn, and a lion for Low. In the medieval period in Germany, the physician is represented on the tombstone by a lion holding a sword. In some communities in Germany, tombstones bore the figures of dragons and bears.

In excavations at Merissa near Beit Jiyrin (3rd century B.C.E.) there was found a grave of Apolophanes, the head of the Zidonian community at Merissa, and on the stucco is painted a cock, the guardian against evil spirits.

In addition, animal decorations have been found on paper cuts traced to Syria, Iraq, and North Africa. A variety of motifs are found, such as a menorah with a seven-branched candelabra, two tablets of the Decalogue at the top decorated with a crown, a *Magen David*, and an eagle. Around them, there are motifs from the animal world, plant life, and geometric forms. The most frequently appearing animals are lions, deer, eagles, tigers, and those that possess a symbolic value and are attributed with special qualities mentioned in the Bible.

Names

Many personal names are linked to the animal world. Some scholars are of the opinion that this adheres to the totem theory, but this cannot be substantiated as the Torah forbids all forms of idolatry. In antiquity, we lived in the open surroundings of the countryside, with its fields, vineyards, and the like. The Bible often uses similes and metaphors of the plant world. It should therefore come as no surprise that in addition to flora, the world of fauna, visible everywhere, should leave its mark on a civilization that saw everything created by God.

In *Avot*, Judah ben Temah says, "Be strong as a lion to do the will of God." Why not go a step further and name a child after the lion, trusting that he will grow up to do the will of God? In Genesis, chapter 49, Jacob uses the animal world to characterize his sons.

Perhaps the answer is in Genesis 2:20. God brings all creatures before Adam to give them names. The traditional explanation for this action is that man could not find one creature who could share his responsibilities and duties until woman was created.

It is possible that, as God brought all creatures before Adam to give them names, man subsequently followed the advice of God and perpetuated these names of animals by passing them on to their children. It is true that animals were not suitable to share the life of man; however, as compensation, man used their names. In this manner, man justified the naming of animals in accordance with the wish of God.

The classical texts in the Bible where important people are compared to animals are Genesis 49 and Deuteronomy 33. Five of the sons of Jacob are given such names–Judah is called a lion's cub, Issachar, a sturdy ass, Dan, a viper or horned serpent, Gad, a rampant lion, Benjamin, a wolf, and Joseph, a firstling bullock.

According to some authorities, we can trace the ox as a name given to a chief (Genesis 49:6). Simeon and Levi hamstrung an ox, which can refer to the chief whose son raped their sister Dinah. Similarly, the "Lords of Edom" are spoken of as "wild oxen," "mighty bulls," and "young steers" (Isaiah 34:7). God Himself is referred to as "the mighty bull of Jacob" (Genesis 49:24, Numbers 23:22).

God is said to have the horn of a wild ox (the Torah refers to God metaphorically). The prophet Hosea metaphorically refers to the decline of Ephraim, a kingdom of northern Israel, by different animal names (Hosea 8:9).

These biblical names are linked to the animal world:

Rachel–ewe lamb–wife of Jacob.

Judah–son of Jacob. Jacob characterizes him as a lion (Genesis 49:9). The lion, being the most majestic of creatures, is the symbol of monarchy. The family of Judah were the rulers of the people of Israel.

Dan–son of Jacob. Jacob characterizes him as a viper or horned serpent (Genesis 49:17). The commentaries explain that the snake is a murderous creature. Throughout the travels of the people of Israel, the tribe of Dan always took up the rear, and it was their duty to wage war with, and to repel, all attacks from other nations, which invariably attacked from the rear (see Rashbam).

Naphtali–son of Jacob. Jacob characterizes him as a hind (Genesis 49:21). Rashi explains that the fruits in Naphtali's portion ripened exceptionally quickly, just as the hind runs at an exceptional speed.

Issachar–son of Jacob. Jacob characterizes him as a sturdy ass (Genesis 49:14). Rashi explains that Issachar, who spent his days and nights immersed in Torah study, is compared to a sturdy ass who can carry a heavy yoke. He carried the heavy yoke of Torah.

Joseph–son of Jacob. Jacob refers to him as a firstling bullock (Genesis 49:22).

Benjamin–youngest son of Jacob. He is characterized as a wolf (Genesis 49:27). Rashi explains that Jacob was prophesying that Benjamin's descendants will be rapacious in the time to come, as we see in Judges 21:21: "and catch you every man his wife" (the story of the concubine of Gibeah). Rashi states further that Jacob was also prophesying concerning Saul, a descendant of Benjamin and a future king of Israel, who vanquished his enemies on all sides.

Hamor–ass–a Hivite prince, the father of Shechem, who afflicted Dinah, the daughter of Jacob (Genesis 34:2).

Tola–worm–the son of Issachar. Just as the silkworm is distinguished for its mouth, with which it spins, so also the men of the tribe of Issachar were distinguished for the wise words of their mouths.

Tziporah–bird–daughter of Jethro and wife of Moses (Genesis 2:21). Tanchuma explains that she was called bird because she purified her father's house of idols, just as a bird purifies a leper of his uncleanliness. Another explanation offered by Tanchuma is that when Jethro's daughters returned after being saved at the well by Moses from the other shepherds, Jethro said to them, "And where is he?" Tanchuma tells us that Tziporah went out as nimbly as a bird and fetched Moses, hence her name (see *Yalkut*, Genesis 2:21).

Gazez–sheep shearer–a son of a concubine of Caleb (1 Chronicles 2:46).

Saraph–kind of serpent (1 Chronicles 4:22). Yalkut actually gives a different explanation and says that *saraph* is the root of the Hebrew word *sereiphah*, "burning." The reason for the name was that this particular person (Saraph) had prepared himself to be burned. He was a descendant of Judah.

Shual–fox–a descendant of Asher (1 Chronicles 7:36).

Shefufan–a kind of serpent–descendant of Benjamin (1 Chronicles 8:5).

Arad–wild ass–descendant of Benjamin (1 Chronicles 8:15).

Hazir–swine–from the Levite family (1 Chronicles 24:15, see also Nehemiah 10:21).

Pirom–wild ass–Canaanite king (Joshua 10:3).

Eglon–from the root word *egel*, meaning "calf"–king of Moab (Judges 3:12).

Yael–mountain goat–a Kenite woman (Judges 4:17).

Orev–raven–a Midianite prince (Judges 7:25).

Zeev–wolf–a Midianite prince (Judges 7:25).

Palti ben Laish–lion (1 Samuel 25:44).

Eglah–calf–wife of King David (2 Samuel 3:2). The Talmud in *Sanhedrin* 21a tells us that this wife was Michal, whom David called Eglah, an endearing name meaning "little calf," because he loved her. She alone, of the six wives of David mentioned, is referred to as "David's wife." Tradition says that she was his first wife.

Zibia–gazelle–Queen mother from Ber Sheva; a descendant of Benjamin (2 Kings 12:2, also 1 Chronicles 8:9).

Jonah–dove (2 Kings 14:25); prophet (Jonah 1:1).

Ha-Aryeh–the lion (2 Kings 15:25). He was so called because he was exceptionally strong (see Radak and *Metsudas David*). Rashi explains that it was not actually the name of a person, but it referred to a golden lion that stood in the king's palace.

Shofon–rock badger–contemporary of Josiah (2 Kings 22:3).

Huldah–weasel–prophetess (2 Kings 22:14).

Aradia–wild ass (Daniel 5:21).

Hagav–locust–family of Nethinim (Ezra 2:45, Nehemiah 7:48).

Yaloh–mountain goat–family of Solomon's servants (Ezra 2:56).

Parosh–flea (Nehemiah 10:15).

Carnio–lamb of Nevo–name of Abraham's grandmother (*Bava Batra* 91a; see Rashi's commentary). The lamb is a pure animal indicating that Abraham was pure in contrast to Haman's grandmother who was named after the raven (*Bava Batra* 91a). Her name indicates the impurity in the family that brought Haman into the world.

Amathlai–name of Abraham's mother (*Bava Batra* 91a). Some etymologists claim that this is a corrupt reproduction of Amathlai-Keren-

Happouch, the name of one of Job's daughters. According to Greek mythology, Zeus was nursed by a goat whose horn overflowed with nature's riches. This, incidentally, is the origin of the cornucopia as a symbol of the horn of plenty; in Hebrew, *Keren Happouch* means "inverted horn."

In the Talmud, in *Sanhedrin* 29b, somebody was nicknamed "the mouse lying on the denarii." Rashi explains that this particular person was very wealthy, but derived no benefit from his money, like a mouse that lies on money but does not benefit at all from it. In Proverbs 5:19 a wife is described as a graceful doe.

The word *nachash* is also found in the Bible as a personal name. It is the name of an Ammonite king (1 Samuel 11:1) and the prince of Judah (Numbers 1:7). It is also found in 2 Samuel 17:27, Exodus 6:23, and Ruth 4:20 and as a name of a city of Judah (1 Chronicles).

Some popular Hebrew names are followed by their Yiddish equivalent: Aryeh-Leib, Tzvi-Hirsch, and Dov-Ber.

Ariel is found in the Prophets where it is exemplified as twin pillars upon which Ariel is founded. Thus Isaiah 29:1 equates Ariel with Jerusalem, and the prophet Ezekiel claims that Ariel is a pseudonym for the Holy Temple (Ezekiel 43:15–16).

However, the name Ariel itself poses a difficulty, for its first part, *Ari* (lion), represents the animal kingdom, whereas *El* (God) represents the Heavenly kingdom. How can two such conflicting ideologies stand side by side? Paradoxically this is the purpose and the function of life. For we are not paragons of righteousness or virtue; we are an amalgam of good and bad, right and wrong, holy and profane. The Almighty has implanted within us both the *Yetzer Tov* (the good inclination) and the *Yetzer Hara* (the evil inclination), and it is our duty and obligation to see that the *Yetzer Hara* enjoys a very low profile and is downgraded while the *Yetzer Tov* should be upgraded and enjoy a noble and exalted profile.

Veterinary Surgery

The rabbis of old were very familiar with some forms of veterinary surgery, for *shechitah* (Jewish ritual slaughter) demands from the *shochet*, who is not a veterinarian, a profound knowledge of the anatomy of the animal.

The process of *shechitah* has two steps. First the *shochet* must locate the precise area in the throat of the animal in which to administer the "cut" by which he severs the jugular vein. This cut then must be followed by *bedikah*, an internal examination that helps the *shochet* detect any deficiencies or abnormalities that may invalidate the *shechitah*.

In addition, the Torah prescribes that all animals offered on the altar should be free from any blemish (Leviticus 22:20). The Mishnah enumerates a list of blemishes (*Bechorot* 5:1–12, 6:11). In addition, meat used for food must come from animals free from lesions.

There are several specific cases in rabbinic literature that refer to the art of healing of animals. The Midrash informs us that Jubal, the son of Lemech, was a veterinary surgeon (*Midrash HaGodel* on Genesis 4:20) and that he also taught veterinary medicine. The basic elements of such medicine were practiced by shepherds and priests. Shepherds are included in the warning issued by the prophet Ezekiel: "Say unto them, 'woe unto the shepherds of Israel'. . . . The weak have not strengthened, neither have you healed that which was sick" (34:2).

During the existence of the Temple, a specialist with the knowledge of a veterinarian was called from time to time to examine all the animals (*Bechorot* 36b).

In Alexandria they used to remove the womb from all the cows and pigs (*Bechorot* 28b)

In the time of the Talmud, there was a veterinarian by the name of Illa who used to accept money for his services (*Bechorot* 29a).

Simon ben Chalafta was an expert veterinarian (*Chullin* 57b), and in the field of experimentation he affixed an artificial leg to a hen (*Chullin* 57a).

The Talmud reports that honey can ease the soreness of the camel (*Bava Metzia* 38b).

We learn from *Mo'ed Katan* 10b that Rav allowed the bleeding of cattle during the festival week.

From *Shabbat* 54b we learn that the rabbis suggest that the legs of the ass be strapped in order to prevent the legs from knocking each other.

In the time of the Talmud, the sages were familiar with cesarean birth, as we learn from Mishnah *Parah* 2:3, "One that is born from the side."

The rabbis refer to a disease caused by a mad dog (*Yoma* 83b).

See *Shabbat* 54b concerning the healing of sheep after shearing; healing of wounds on the camel (is discussed) in *Bava Metzia* 38a.

In *Bechorot* 40a Rabbi Akiva discusses testing the testicles for blemishes.

Ben Sira (200 B.C.E.) writes that the physician and his art of healing are God's creation.

The Lord has cultivated drugs from the soil (*Ben Sira* 38:2,4).

Rav Ashi (4th century C.E.), when confronted with an inquest on *treifah*, would assemble all the butchers of Matha Mehasia (Mesopotamia) for a consultation (*Sanhedrin* 7b).

The Talmud informs us that hysterectomy was known to the rabbis 2,000 years ago. In the days of Rabbi Tarfon, a case was presented to the sages, who ruled that the feeding of dogs with a cow whose womb was removed was permitted (*Bechorot* 4:4).

Animal treatment by drugs was known, and it was permissible to obtain such drugs from pagans (*Avodah Zarah* 2:2).

Shoeing of horses was practiced, and the shoes were made of metal or cork (*Shabbat* 59a).

Obstetrics was a part of a shepherd's work (*Chullin* 4:3).

Post-talmudic Literature

In the *Kuzari*, Yehuda HaLevi (11th century) refers specifically to the problem of animal suffering:

> See how wonderfully conceived is the nature of the creatures, how many marvelous gifts they possess which show forth the intention of an all-wise Creator, and the will of an omniscient all-powerful Being. He has endowed the small and the great with all necessary internal and external senses and limbs. He gave them organs corresponding to their instincts. He gave the hare and the stag the means of flight required by their timid nature; endowed the lion with ferocity and the instruments for robbing and tearing. He who considers the formation, use, and relation of the limbs to the animal instinct, sees wisdom in them and so perfect an arrangement that no doubt or uncertainty can remain in his soul concerning the justice of the Creator. When an evil thought suggests that there is injustice in the circumstance that the hare falls prey to the lion or the wolf, and the fly to the spider, reason warns him as follows: How can I charge the All-Wise with injustice when I am convinced of His justice, and that injustice is quite out of the question? If the lion's pursuit of the hare and the spider of the fly were mere accidents, I should have to assert the necessity of accident. I see, however, that the wise and just Manager of the world equipped the lion with the means for hunting, with ferocity, strength, teeth, and claws; that He furnished the spider with cunning and taught it to weave a net which it constructs without having learnt to do so; how he

> equipped it with the instruments required and appointed the fly as its food, just as many fish serve other fish for food. Can I say aught but that this is the fruit of a wisdom which I am unable to grasp, and that I must submit to Him who is called: "The Rock whose doing is perfect" (Deuteronomy 32:4).[1]

Maimonides (12th century) was convinced that everything in God's creation is for the use of man; leaves with deadly poison have curative value. In *The Guide for the Perplexed* he writes:

> Wherever you find animals or plants which are unsuitable for food and are useless according to your thinking, know that this is due to the weakness of our intellect. It is impossible for any living creature from the elephant to the worm to be void of all utility for man. The proof of this is that in every generation there are discovered by us important uses for herbs and various kinds of fruits which were unknown to our predecessors. Through experimentation by successive generations that which is unknown becomes known. If a human being is deprived of his intellectual faculties and possesses only vitality, he becomes like a beast and is soon lost (*Guide for the Perplexed* 1:72).

The *Sefer Chasidim* by Rabbi Judah HeChasid (13th century) is full of restrictive measures to be adopted by those who wish to be forgiven for their sins. Yet, at the same time, it contains several humane precepts that the author advises the pious man to follow. A few points regarding the treatment of animals are provided here:

"Man should not withhold from his animals any of the gratitude it might have earned."

"The man who is cruel to animals will have to answer for it on the Day of Judgment, and the driver will be punished for applying the spur too often."

Moses ben Eliezer HaCohen, in his *Little Book of Saints*, writes, "Be careful to feed the poultry in thy house before thou takest thy meal."

Judah ben Samuel HeChasid of Regensburg, in *Sefer Chasidim*, writes: "Refrain thy kindness and thy mercy from nothing which the

Holy One, Blessed Be He, created in this world. Never inflict pain on any animal, be it bird or insect, nor throw stones at a dog or a cat, nor kill flies or wasps" (*Sefer Chasidim* 13a). The saints were extremely pious and gave the impression that they were otherworldly, concerned only with prayer and study of the celestial things above, and uninterested in the mundane affairs of life. Obviously, however, saintliness and piety must include an understanding and kindly approach to the dumb animals. This too is part of the holiness code.

In a long extract dealing with a variety of animals, Joseph Albo (15th century) displays his vast knowledge of the animal kingdom. For instance, he writes:

> The horned animal eats grass; the material intended for the teeth was used up in the formation of the horns and nature was not able to make teeth for them in the upper jaw, as a consequence of which they cannot masticate their food properly. Hence nature gave them the power to chew the cud in order that they might complete in the second mastication what was left undone in the first. Again, those animals that can find food readily digest their food quickly, as for example fish and fowl, as the rabbis say: "In the case of fish and fowl, their food is digested in the time it takes them to be consumed when thrown into the fire." Beasts and birds of prey, which do not feed on plants, are provided by nature with organs which enable them to feed on what they catch, and with poison in their claws which they inject into the animal they kill, and which cooks the food and grinds it.[2]

Albo described these animals as well. The snake is able to inject poison into its enemies by means of two hollow teeth resembling a hypodermic needle. The squid is a long, slender sea mollusk having ten arms, which squirts out an inky barrage to conceal it from its pursuers. The eel is a reptile that generates an electric current that gives a shock to its enemies. The firefly is an insect that is able to turn on and off a light that it bears in its translucent belly. And he concludes, If we reflect upon the formation of the animal species and the constitution of their organs, we find that the Creator has exhibited wonderful care in ordering their affairs and their every need in a remarkable manner.[3]

The Maharal (16th century) encouraged people to feed the fish in tanks kept at home during festivals.

We may learn that a man ought to study everything that will enable him to understand the essential nature of the world. One is obligated to do so for everything is God's work. It should be understood in its entirety, and through it one recognize one's Creator.[4]

In *Sefer HaMinhagim*, written by Rabbi Moses Isserles (16th century), it is stated, "It is customary to say to a person putting on a new garment, 'May you wear it out and get a new one.' Some write that one ought not say it to a person putting on new shoes or clothing made from the skins of animals, for it would then seem as though the animal was being killed to make a garment, and it is written, 'His tender mercies over all His works.' " Isserles also wrote, "One who slaughtered an animal for the first time ought to recite a blessing, 'Who has kept us alive,' when he covers the blood of an animal, not when he slaughtered it, for he is injuring a living thing. This accounts for the prohibition against wearing shoes on Yom Kippur, for how can one put on garments for which it is necessary to kill animals on a day of compassion when it is written, 'and His tender mercies.' "

God created animals with innate powers of self-preservation and means to combat their enemies. For example, a skunk is able to shoot a suffocating stench against its enemies. A porcupine can present javelins against its attackers.

Rabbi Israel Salanter (19th century) taught, "Man may be compared to a bird. It is within the power of the bird to ascend even higher on condition that it continues to flap its wings without cessation. If it should stop flapping its wings even for a moment it would fall into the abyss" (*Tenuat HaMusar*, p. 269).

Once Rabbi Salanter was going to the synagogue for the *Kol Nidre* service. On the way he encountered an animal belonging to a gentile, which appeared to be lost. He saw that the animal was in distress and took the trouble to lead it home over stones and rocks and through fields and gardens. Meanwhile his congregation was waiting for him. When he did not come they went out to look for him and found him trying to lead the animal into its master's stable.

A child who is not reprimanded for laughing at the agonizing outcry of an animal will subsequently grow up to disregard his parents' call in distress.[5]

In ancient Egypt, God's attributes, qualities, and acts were personified by separate divinities that were represented by the grossest and most monstrous symbolism. The bull, the cow, the cat, the ape, crocodile, hippopotamus, hawk, ibis, scarabeous, and others were each emblems of a divine personage. "If you enter temples," says Clemens of Alexandria, "a priest advances with a solemn air, singing a hymn in the Egyptian language; he raises the veil a little to let you see the god, and then what do you see? A cat, a crocodile, a snake or some other animal. The god of the Egyptians appears. It is but a wild beast wallowing on a purple carpet."[6]

A. Marmorstein, in his book *Studies in Jewish Theology*, writes: "Tertulian and others ridiculed by Marcion mention the creator of small stupid insects. It must be in view of these agnostics and other enemies of the Bible that some preachers enunciated an often repeated saying: 'Even those things which seem to you superfluous in the creation, e.g. flies, etc, are part of the creation and act as God's messengers.' "[7]

In *Guardians of Our Heritage* Leo Jung writes:

> Enjoying as we did, the little birds which flew back and forth from our hill to the Mosque of Omar, the multi-colored butterflies, the ants on their hills, we were naturally bound to ask ourselves whether or not they knew that God had created them. We pondered upon it a great deal and the conclusion made us very sad. The most sorrowful distinction between animals and man is their not knowing that God created them and that they could lean on him as we do. We ought to be doubly kind to them to make up for their unfortunate ignorance. I had a little cat which I cherished and with which I shared my food. This single act is not sufficient. We must spread a larger quantity of grains for little birds and be attentive to the needs of some other animals in the

yard. Would this Mitzvah be among the 613, or did we increase the number by one?[8]

From an address by Nina Adlerblum entitled "Memories of Childhood, an Approach to Jewish Philosophy," we quote:

> Another perturbance about the soul came to us through stories from the Hassidim. Sitting on their stoop, we heard a long discussion on how some souls would travel from one body to another and assume even the shape of an animal. "If my soul has to travel, I prefer that it should enter a cow rather than into the body of an *apikoros* (nonbeliever)," I exclaimed tremblingly. But we reassured one another that if the soul has to enter into different bodies, it could only go into better ones. In this manner, little by little, the world becomes a perfect one.[9]

In the same essay—on Purim—we read:

> We also discussed the animals in Noah's Ark and we were glad they cheered him up on the lonely waters. They must have been as friendly as the colorful birds nested on the roof of the Hassidic synagogue. They would eat out of our hands and we could almost recognize them by their distinctive feathers and by their age. A tiny, still featherless one hatched out of an egg and was greeted with Mazal Tov and handshaking. We should have liked to have given them Jewish names since they were born on a Jewish terrace, but names are given at the synagogue. On Purim, we gave the birds crumbs from our Hamen Taschen and they liked them.[10]

The granddaughter of Rabbi Isaac Elchanan Spector tells of occasions when the rabbi took her in the summertime to a resort on a mountain near Kovno. "In the resort I was greatly impressed by his cordial relations with the domestic animals. The cat, for instance, was sitting in a chair behind the back of grandfather and was eating directly from his hand. His love for nature was limitless. He greatly enjoyed flowers and would become intoxicated by the beauty of a tree and the grand view of a sunset."[11]

Rabbi Abraham Isaac Kook (1865–1935) was very tolerant of the irreligious. To critics of this tolerant attitude he would reply by referring to the law of the firstling of the ass. Of all the unclean animals, the ass was the only one to have its firstborn sanctified, despite the fact that it lacked both inward and outward characteristics that distinguish unclean from clean beasts. The reason given by the rabbis for this sanctification was that the ass helped carry the luggage of the Israelites when they left Egypt and made their way to the Holy Land. Surely, even assuming that these irreligious *chalutzim*, pioneers settling in Palestine, are, as is maintained, bereft of all Jewish piety, both internal and external, their endeavors in assisting the Jewish people out of *galut* (exile) and in rebuilding the Holy Land stamped them with the distinction of holiness.

Of Rabbi Kook, it is said: He loved every manifestation of natural life. His love extended even to animals and birds and embraced the rest of creation. He saw the whole of nature constantly rising toward that Godliness that will find its consummation in divine messianic regeneration.[12]

J. Wohlgemuth in his article "Consideration for the Animal in Judaism," stresses the vital need for the urbanized Jew to return to nature.

The miracle of birth and maturation of animals (Job 39:1–4). The passion of certain animals for freedom and their refusal to submit to the domination of man in contrast to the readiness of other creatures to allow their domination (Job 39:5–12). The success of creatures who have seemingly been deprived of wisdom by their creator to perpetuate themselves and thrive (Job 39:13–17). The martial glory of the horse (Job 39:19–25). The majestic soaring of mighty birds and their distance, conquering vision (Job 39:26–30). The monsters with whom man struggles and those whom he cannot even approach (Job 40:15, 41:26). All these cannot be graduates of blind forces in a meaningless world.[13]

The lower animals were allowed to participate in the privileges of the Sabbath in common with their owners. With the exception of the Jewish code, it does not seem that the useful animal ever obtained benefit from any legal enactments. The principle of humanity to beasts of burden was never assured on the basis of legislation in any of the national codes of the ancient world. The relationship between man and his cattle can, according to Isaiah, be compared with the relationship between father and son. The man has to covenant with his cattle just as he has a covenant with his earth. Just as he is not permitted to exhaust the earth, so he must not weaken the lives of the cattle. Of the slave it is said that it is in order that he may be ensouled. The same might be said of cattle.[14]

Perek Shirah

Perek Shirah is an amazing little book, the author of which is unknown. It was probably written in the tenth century though some traces of it are found in the Talmud. This book has three parts: heaven and earth, plants, and animals. However all three parts share one common aim; they all owe their allegiance to their Creator and sing songs of praise unto Him. Hence, the title of *Perek Shirah*, a Chapter of Song.

The style of the book is quite simple and with the exception of the *tarnegol* (the cock), each animal sings a song of praise consisting of one or two appropriate verses from the *Tanach*.

At first glance the notion that animals sing praises unto God may seem strange, but upon further scrutiny we discover that this idea is well rooted in the Bible and in rabbinic tradition.

Beginning with the Bible, we read in Psalm 104:21 the following: "The young lions roar after their prey, and seek their food from God." Here we learn that animals petition the Almighty to provide them with food. Their song of prayer is expressed in the roar that emanates from their mouths. Now that we have substantiated our thesis that animals can pray, we proceed to another text in the Bible from which we learn that animals can sing.

In 1 Samuel 6:12, referring to the cart that was drawn by cows in which the Holy Ark was carried from the country of the Philistines, the

Rabbis ask, "What is the meaning of the word *yisharna*" (took the straight way)? The Rabbis trace the root to the word *shir* (song), and Rabbi Jonathan in the name of Rabbi Meir therefore translated it, "And they rendered song." Rabbi Zutra Bar Tobiah said in the name of Rab, "They directed their faces toward the Ark and rendered song. What did they sing? They sang, 'Then sang Moses and the children of Israel' " (*Avodah Zarah* 24b).

We now turn to the prayer book with its rich resources and meaningful texts. Commenting on the last verse of Psalm 150, "*Kol haneshamah* . . . Let everything that hath breath praise the Lord," the Midrash says, "*Kol neshamah* . . . for every breath that one takes one must praise God" (*Deuteronomy Rabbah* 5:2, *Tanchuma*, chapter 9). As animals also breathe, this responsibility for praising God must also apply to the animal kingdom.

The above comment may also be extended to apply to the opening words of *Nishmat*, which is recited on Sabbath and festival mornings. Instead of its usual translation as "soul," it can be translated as "breath," so that the prayer reads, "The breath of all life shall bless Thy name." In this way, the animal world can be included in blessing God: "And the spirit of all flesh shall continually glorify and exalt Thy memorial." Compare Psalm 136:25: "Who giveth food to all flesh."

Following the *Nishmat* prayer, we read, "For such is the duty of all creatures (*hayetzurim*) in thy presence . . . to thank, praise, laud, glorify, extol, honor, bless, exalt, and adore." The operative word in this prayer is *hayetzurim*. If the word *Adam* (man) or similar expression was used, it would obviously refer to the human race, but *hayetzurim* has a wider connotation. There is talmudic evidence for this wider meaning (see *Niddah* 22b where the rabbis compare Genesis 2:19, "and God formed [*Vayetzer*] every beast of the field," with *hayetzurim*).

Zemirot, the songs sung at the table on the Sabbath, usually begin with "*Yah Ribon*," and in the second paragraph we recite the following words, "Early and late to Thee our praises ring, Giver of life to every living thing, beast of the field, and birds that heavenward wing." These words are the very essence of *Perek Shirah*.

Yet, another passage in the prayer book warrants our attention. In the Rosh Hashanah and Yom Kippur *Amidah*, immediately after the introductory blessings, we read, beginning with the words "*Uvechein tein pachdecha*"–"Now therefore, O Lord our God, impose Thine awe upon all Thy works, and Thy dread upon all that Thou hast created, that all may fear Thee and all creatures prostrate themselves before Thee, that they may all form a single bond to do Thy will with a perfect heart, even as we know, O Lord our God, that dominion is Thine, strength is in Thine hand, and might in Thine right hand, and Thy name is to be feared above all that Thou hast created."

This prayer, with its repetition of *kol*, all or everything, highlights the ethical concept of universalism that permeates Jewish thought. This universalistic ideal signifies that on Rosh Hashanah and Yom Kippur we celebrate the anniversary both of the human race and the animal kingdom. Universalism is also reflected in the rabbinic teaching that the world was created on Rosh Hashanah.

It is interesting to point out that the Midrash categorically states that the *Yatush*, "the mosquito," was created before the birth of man (*Genesis Rabbah* 5). Furthermore the Torah mentions that God created (*bara*) the great sea monster (Genesis 1:21).

With this background in mind, we may paraphrase the words of the second verse of the above prayer as follows: All creatures shall worship Thee and sing songs of praise unto Thee. This indeed is the message of *Perek Shirah*.

We conclude with an appreciation of *Perek Shirah* by Joseph Albo in his *Sefer Ikkarim*: "The Rabbis allude to *Perek Shirah* (gaonic compilation containing praises of God sung by heavenly and earthly bodies and by plants and animals) and ask, 'What do creeping things say?' and reply, 'May the glory of the Lord endure forever; let the Lord rejoice in His works' " (Psalm 104:31).[1]

Therefore, although it may not be becoming to God's dignity and His excellence and perfection to rejoice in such inferior creatures, nevertheless, we may speak metaphorically of His rejoicing in them because they are His work. And the meaning of "May the Glory of the Lord endure forever" is that without slighting in any way God's glory, which is eternal, we may say nevertheless that these creatures, born of putrefaction, low and insignificant as they are, having no permanence either as a species or as individuals, give God cause for rejoicing because His wisdom is in them, which are His work.

THE ANIMAL KINGDOM IN JEWISH THOUGHT

We now turn to a detailed study of sixty-five different species of animals, all found in the Bible and reviewed in alphabetical order.

Before analyzing the various species of animals, it is necessary to examine the generic term "animal," which is expressed in Hebrew as *chayah* or *behemah*.

Chayah is derived from *chai*–"living (creature)"–and is often used in conjunction with the word *raah*–"evil"–to refer to wild beasts (see Genesis 37:20, Leviticus 26:6). Marcus Jastrow, in his *Dictionary of Targum, Talmud and Midrash*, defines *chayah* as "a beast of chase" as opposed to *behamah*, which is linked to the root *beham*, meaning "dumb," and is generally used in reference to domestic animals (mostly of the horned type).

Another term occasionally used is *beir* (Exodus 22:4, Numbers 20:4), which is translated as "grazing animals or cattle." It is interesting to note that Rashi in *Bava Kamma* 17b on "*Beiro*" suggests it to be the translation of *behamah*.

Yet, another generic term for animals is *baalei chayim*–literally "masters or owners of life" (see *Sanhedrin* 57a, *Pesachim* 98a).

Ant

It is estimated that there are 100 species of the ant, and the Hebrew *nemalah* is probably a generic term for all of them. Authorities differ about the origin of the word *nemalah*. Shoshan suggests that the name is derived from the root *malal*–"to rub or scrape."[1] In rabbinic Hebrew *malal* is used for ears of corn and may therefore refer to ants who gather the wheat in the harvest. Jastrow in his talmudic dictionary surmises that *nemalah* is a contraction of *ne'amalah* and is derived from the root *amail*–"to toil."[2] As we shall see, the ant is famed for her toil and industry. However, the name is more likely derived from *mool*–"to circumcise or cut" (compare Genesis 17:11, *unemaltem*–"and you shall cut"). This may well refer to the action of the ants who with their strong nippers cut or bite off the germs of the grains to prevent them from sprouting and so preserve them from rotting (Gershon ben Shlomo, in *Shaar HaShamayim*).

The ant has been popularized in two classical texts in Proverbs 6:6 and 30:25, where the ant is highlighted as an example of dexterity and industry. "Go to the ant, thou sluggard, consider her ways and be wise: which having no chief overseer or ruler provides her bread in the summer and gathers her food in the harvest." King Solomon was deeply concerned about the dangers of slothfulness, and the lazy person is satirized in Proverbs 19:24 and 26:14–16. In our text the author does not offer destructive criticism, but rather advises the sluggard to adopt a constructive course of action–Go to the ant and be wise. The Rabbis elaborate on the theme of idleness and affirm that it leads to lewdness and mental instability (*Ketubot* 59b).

The commandment to observe the Sabbath is preceded by the injunction "Six days shalt thou labor and do all thy work" (Exodus 20:9). The dignity of labor is very evident in Jewish teaching. The antithesis to *atzilut* (laziness) is *zerizut* (enthusiasm, alertness, and activity), qualities that are characteristic of the ants that collect their food in the summer when the grain is plentiful and lay up stocks for the winter when they hibernate. This initiative on the part of the ants is all the more praiseworthy because they are self-reliant and finish the task they assign to themselves without supervision; they have "no chief, overseer or ruler."

A second-century talmudic sage, Rabbi Simeon ben Halafta, who was a keen observer of nature and was determined to ascertain by experimentation the veracity of the above scriptural verse, threw a cloak over an ant heap. As ants prefer to be hidden from the sun, one ant appeared. The rabbi marked it so that he would recognize it later. Observing the shadow, the ant hurried back to report what it had seen, and a multitude of ants appeared. At this same moment, the rabbi removed his cloak, and the crowd of ants, believing that they were deceived, fell upon the messenger and killed it. The rabbi argued that if the ants had a king or ruler, he would not have tolerated such a travesty of justice (*Chullin* 57b, *Yalkut Shimoni* on Proverbs 6:6). The absence of a ruler here is considered a vice and not a virtue.

Another characteristic of the ant that should be emphasized is its absolute honesty. The Torah often denounces deception and fraud and lauds those who deal honorably. It is therefore not surprising that we are advised to follow the ways and habits of the ants who enjoy a moral code that exalts uprightness of conduct and is ingrained instinctively in them. The integrity of the ant is highlighted in the famous talmudic dictum that states that if the Torah had not been given, we could have learnt modesty from the cat, honesty from the ant, chastity from the dove, and good manners from the cock (*Eruvin* 100b).

This honorable trait in the ant is illustrated by the same nature-loving Rabbi referred to earlier, R. Simeon ben Halafta. He once saw an ant that had dropped a grain of wheat, and he then observed that many ants passed, smelled the grain, but none would touch it. Eventually its rightful owner returned and picked it up (*Deuteronomy Rabbah* 5:2). We thus see that although indolence may lead to robbery, the ceaseless toil of the ant leads to rectitude.

The second text in the Book of Proverbs relating to ants is as follows: "The ants are a people not strong, yet they provide their food in the summer" (Proverbs 30:25). This verse is preceded by one that states, "There are four things which are little upon the earth but they are exceeding wise."

Here we learn that small creatures (and the ants head the list) also contain elements of greatness. Indeed the ants are designated as "a people," and a people flourish when they are united by bonds of loyalty and considerateness. The Hebrew word for people, *am*, is closely connected with *im*, meaning "with or together." The ants enjoy a togetherness; rarely are they seen alone. They appear in multitudes, and it is common to speak of colonies of ants. In contrast, large animals such as the lion, tiger, and wolf are often alone. They find protection in the power and bodily vigor of their own strength, but the ants are small creatures and flourish in groups and societies. Like human beings, the ants survive because they are united.

The ants too constitute a people because they plan and build their own homes deep in the ground where is found a maze of galleries and subterranean passages methodically and carefully designed and erected with great skill and ability. There they live as a people, a community governed by a distinctive code of honor. The Midrash presents us with a blueprint of their building operations, which reveals the wisdom with which they devised their underground fortresses. Thus, we are informed that there are three stories; ants use the middle one to stock their grain because the top story might be damaged by the rain and the bottom by the dew of the earth (*Yalkut Shimoni* 938). As a people, the ants exemplify great resourcefulness, sagacity, and wisdom.

Yet, the ants are described as a people "not strong." This characterization seemingly contradicts the well-founded assumption that the ants are the strongest creatures on earth since they are reputed to carry two to four times their own weight. To reconcile this seeming contradiction, we contend that the word *az* does not necessarily mean "strong," but rather "fierce or ferocious"; for example, "a nation of fierce countenance" (Deuteronomy 28:50). The lion, for instance, can inspire dread and terror by its massive mane of hair and thunderous roar, but the insignificant ant looks very tame; it is *lo az*, it is not fierce or formidable. Furthermore, large animals violently attack their prey when they are hungry, but the ants will not waste time, they will not wait for pangs of hunger to overtake them; they plan ahead and quietly "provide their food in the summer."

Rabbi Tanchuma observed that the average life span of an ant is six months because it has no bones or sinews, and such creatures cannot live long. The Midrash also informs us that the puny ant needs no more than one and a half grains to nourish it. Why then do ants gather in more than they require? Indeed, Rabbi Simeon ben Yohai reported that in one ant's nest there were 300 bushels of grain. In the words of the Midrash the ants hope that the Almighty may lengthen their lives, and they must not be found wanting, so cautious are they (*Deuteronomy Rabbah* 5:2).

The ant provides us with yet another lesson found in a remarkable Midrash that invokes a large number of witnesses testifying against Israel and warning her of her impending doom. Finally the Midrash invokes the testimony of the ant: "Go to the ant . . . consider her ways and be wise." The Midrash closes with these incisive words: "It was a sufficient humiliation for man that he had to learn from the ant; had he learnt and acted accordingly he would have been sufficiently humbled but he did not learn from the wise ant" (*Yalkut Shimoni* on Joshua 24:22).

The central thought embodied in the above Midrash is reflected in an interesting story about King Solomon and his reputed ability to understand the language of the animal kingdom. In the course of his wanderings Solomon heard an ant issuing orders to others to withdraw and so avoid being crushed by the armies of the king. Thereupon Solomon summoned the ant who informed him that she was the queen of the ants and offered reasons for her orders. Solomon wished to question her but she defiantly refused to answer unless the king took her and placed her on the palm of his hand. He acquiesced and repeated his question, "Is there anyone greater than I am in the world?" asked Solomon. The ant promptly retorted, "Yes, I am." Solomon was taken aback, saying, "How is that possible?" The ant, unperturbed, replied, "Were I not greater than you, God would not have led you here to place me in your hand." Exasperated, Solomon threw her to the ground and exclaimed, "I am Solomon, the son of David." Not to be intimidated, the ant reminded the king of his earthly origin and admonished him to be humble. The king went his way, feeling abashed. In this manner, the wise Solomon was outwitted by the unpretentious ant.[3]

Throughout the ages, Jewish scholars have found the study of the ant enchanting. One talmudic sage was so visibly moved when he beheld the wondrous acts of the ant that he felt impelled to exclaim with the Psalmist: "Thy righteousness is like the mighty mountains: thy judgments are like the great deep; man and beast Thou preserves: O Lord" (Psalm 36:7, *Chullin* 63a).

In the post-talmudic era, a host of scholars have discoursed on the ant. Maimonides, Bahya Ibn Pakudah, Albo, Ralbag, and the Vilna Gaon all have been charmed by the wisdom and foresight of the ant. A summary of the ant's admirable qualities is presented by Albo in the *Ikkarim*:

> If one reflects upon the ant, which is a tiny creature and yet prepares her food in the summer and gathers it in the harvest, he will learn the quality of industry and will understand that God gives food to all flesh and prepares sustenance to all creatures and that He has endowed each one with the instinct of seeking its food at the proper time and from the proper source.
>
> Man therefore, who is endowed with knowledge and understanding, must make efforts to obtain food and not be idle. We learn the wrongfulness of robbery from the ant which never steals or robs or takes anything that belongs to another ant; it will refuse to take what does not belong to it.[4]

Proverbial Saying

Go to the ant, thou sluggard, consider her ways and be wise (Proverbs 6:6).

Ass

In the biblical period, the ass served a dual role; it carried burdens (including pulling the plow), and people regularly rode on them. They served as the main form of transportation in early days.

Although the ass was a beast of burden, it was not to be exploited. Thus we are enjoined, "You shall not plow with an ox and ass together" (Deuteronomy 22:10). Compared with the ox, which is a heavy animal, the ass is slight and normally takes quick, short steps. Therefore, yoking them together would make the burden unequal for both animals and cause pain. This humanitarian approach to the animal is extended even to one's enemy: "If you meet your enemy's ox or his ass going astray, you shall surely bring it back to him again. If you see the ass of him who hates you lying under its burden, you shall forbear to pass by it; you shall surely release it with him" (Exodus 23:4–5). Why should the ass be penalized because of an enmity that exists between two human beings? It is therefore a positive command to render assistance to the animal whether it is lost or is suffering under an excessive load.

Let us now consider another law concerning the ass: "And every firstling of an ass you shall redeem with a lamb" (Exodus 13:13). The ass was an unclean animal and therefore could not be sacrificed. Yet, the ass alone, and no other unclean animal, is mentioned in this law, because the ass was of assistance to the Israelites when they left Egypt. For there was not a single Israelite who did not take with him from

Egypt several asses laden with silver and gold (Rashi based on the *Mechilta* and *Bechorot* 5b). This law teaches us the significance of gratitude. Jewish ethics demand that we should continually show our appreciation for kindnesses we receive. The ass does not possess any of the signs of *kashrut*; it does not chew the cud nor has it a divided hoof. Yet it was considered holy and had to be redeemed with the lamb because it helped the Israelites by carrying their burdens when they left Egypt. Such consideration to the animal world is the highest form of gratitude.

In addition to carrying the physical burdens of man, the ass is figuratively portrayed as carrying the spiritual burdens of Torah. Commenting on Genesis 49:14, "Issachar is a large-boned ass," the Sages affirm that "as the ass carries burdens, so Issachar carries the yoke of the Torah" (*Genesis Rabbah* 99:40). Both the ass and the tribe of Issachar were strong, hard working, and resolute and were destined to serve mankind in different spheres of life.

An allusion to the role of Issachar as a devout and accomplished student of Torah is found in 1 Chronicles 12:33: "And of the children of Issachar, men that had understanding of the times, to know what Israel ought to do; the heads of them were two hundred." In another source we learn that the two hundred chiefs mentioned above were the leaders of the Sanhedrin and their decisions were followed without question (*Genesis Rabbah* 72:5, 98:12).

We shall now discuss riding or traveling, for which the ass was extensively used. It is noteworthy that, although the ass was a beast of burden, it was cherished by princes and leaders of men. Thus of Abdon, the son of Hillel the Pirathonite, the Bible says, "He had forty sons and thirty grandsons that rode on three score and ten ass-colts" (Judges 12:14); and Jair (Judges 10:4) had thirty sons who rode on thirty ass-colts.

There were two outstanding instances in the Bible in which influential people rode on asses. One is the speaking ass on which Balaam rode, recorded in Numbers, chapter 22. Whatever interpretation we follow, the message of this narrative is patently clear and is forcibly presented by Maimonides:

> There is a rule laid down by our sages that it is directly prohibited in the Law to cause pain to an animal and it is based on the words: "Wherefore have you smitten your ass?" (Numbers 22:32). But the object of this rule is to make us perfect, that we should not assume cruel habits and that we should not uselessly cause suffering to others; on the contrary we should be prepared to show pity and mercy to all living creatures, except when necessity demands the contrary.[1]

It is significant that the ass was chosen by God to be the vehicle of expression of the rights and privileges of the animal kingdom. Man is distinguished from the animal because he is endowed by God with speech. In this story, the ass takes on the mantle of man.

Another example is found in the Book of Zechariah: "Behold your king comes to you, he is triumphant and victorious, lowly and riding upon an ass" (Zechariah 9:9). In the words of Rashi, "This can only refer to King Messiah of whom it is said, 'And his dominion shall be from sea to sea,' since we do not find any ruler with such wide dominion during the days of the Second Temple."

In this prophetic vision of the Messianic Era, the ass is contrasted with the horse. The latter is the beast of war galloping triumphantly from the field of battle. The former is the quiet, reserved animal symbolizing peace, carrying the burdens of mankind willingly and unbegrudgingly.

One additional aspect of the domesticated ass needs to be discussed and clarified. It is universally acknowledged that the ass is a foolish animal, but how do we reconcile this characterization with the foregoing accounts? The Mishnah in *Pirkei Avot* 5:9 declares that the mouth of the ass was one of the ten miracles created before the Sabbath. However, we cannot envisage the Almighty performing a miracle through the medium of a foolish animal, nor can we look forward to the future King Messiah riding on a stupid ass. It is true that in the Talmud several proverbial sayings speak of the ass in a derisive fashion, but they have no scriptural authority. There is a passage in the *Tanach* that treats the ass in an uncomplimentary manner. It is possible that the derogatory connotation given to the ass crept into Jewish thought during the Greek or Roman period. We know that Jerome (4th century) translated the Bible into Latin, but did not resort to the Greek translation; he preferred to use the original Hebrew text. When his critics attacked him for doing so, he retorted by calling them "two-legged asses."

There is, however, one passage in the Bible that calls for examination. In the story of the *Akedah* (Genesis 22:5), Abraham says to his two youths, "Abide you here with the ass and I and the lad (Isaac) will go yonder." The Talmud (*Yevamot* 62a) infers from this verse that certain people are to be compared to the ass for lack of status. However, we contend that this verse is actually a simple observation. The two young men were not inspired to the same degree as Abraham and Isaac. They could not reach the celestial heights of Moriah; they were therefore asked to stay behind with the ass because their task was completed. In addition to the above midrashic interpretation, the halachic implication in the Talmud is very clear: It deals with the legal status of a slave, and the Rabbis postulate that as a slave is the chattel of the master, so is the ass. There is no hint here of alleged stupidity. Indeed one writer goes to the other extreme and is of the opinion that the word *ass* is a laudatory term meaning "chieftain or head of a tribe."[2]

At any rate, we prefer to direct our attention to the story recounted in *The Fathers according to Rabbi Nathan*, chapter 8:

Once the ass of Rabbi Chanina ben Dosa was stolen by brigands. They tied it up in a yard and put before it straw, barley, and water, but it would not eat or drink. They said, "Why should we allow it to die and befoul our yard?" So they opened the gate before it and drove it out. It walked along braying until it reached the house of Rabbi Chanina ben Dosa. When it arrived, the rabbi's son heard its voice and said to his father, "This sounds like our beast." Said the rabbi, "My son, open the door to it for it has almost died from hunger." Immediately the son placed before it straw, barley, and water and it ate and drank. Therefore it was said, "Even as the righteous of old were saintly, so were the beasts saintly like their masters."

We cannot conclude this review without mentioning the three wild species of the ass found in *Tanach*: *ayir*, *pere*, and *arod*. *Ayir* and *pere* occur in one verse, Job 11:12, whereas *pere* and *arod* are found in Job 39:5. Samson Raphael Hirsch, in his Bible commentary, designates *ayir* as a lively mettlesome young donkey and *pere* as a wild animal that wishes to be free from the human yoke. *Arod*, which is considered an Aramaic loan word, is defined in Mishnah *Kilayim* 8:6 as belonging to the class of beasts of chase. It is an uncontrolled animal that cannot be tamed.

Proverbial Saying

Sweet to the donkey is his bray, like to the song bird its song (Joseph Caspi on Proverbs 24:16).

Bear

The Hebrew word for bear, *dov*, is derived from the root *davav*, "to move gently or to glide." The B. D. B. Hebrew Lexicon[1] suggests that the bear is an animal that moves gently and slowly; hence, *dov*.

Today the bear is extinct in Israel, apart from those that are in the Biblical Zoo in Jerusalem. However, it was present in biblical days, and is mentioned several times in the *Tanach*. Second to the lion, the bear is a most ferocious and dangerous animal: "As if a man did flee from a lion and a bear did meet him" (Amos 5:19).

Shepherds were afraid of losing their flock to bears, as we learn from reading about the prowess of David. "And David said unto Saul, 'Thy servant kept his father's sheep and when there came a lion or a bear and took a lamb out of the flock I went out after him and smote him and delivered it out of its mouth; and when he arose against me I caught him by his beard and smote and slew him' " (1 Samuel 17:34).

Because of the ferocity of the bear we can understand the metaphor used by Isaiah: "the cow and the bear feeding together" (Isaiah 2:7). This is symbolic of the vision of profound peace of messianic times.

In the apocalypse of Daniel (7:5) because of its greediness, the bear is a symbol of the Median empire, which is greedy for lands.

A characteristic of the mother bear is her close attachment to her young and her dread of losing them: "They are mighty men and embittered in their minds, as a bear robbed of her whelps in the field" (2 Samuel 17:8; compare Proverbs 17:12; Hosea 13:8).

A sad incident is recorded in 2 Kings 2:23–24 when Elisha cursed the young persons who mocked him: "And there came forth two she-bears out of the wood and tore 42 children of them." It is worth noting that Elisha was punished with sickness for having "incited the bears against the children" (*Sanhedrin* 107b).

In midrashic literature the word *dov* allegorically meant "temptation." In *Genesis Rabbah* 87:3, we learn that Joseph had to pass through different stages before he could refine his character. When he realized he was to rule over Egypt, he pampered himself with food and drink and began to curl his hair. Such conduct did not find favor in the eyes of God, who castigated him with these words, "I shall arouse against you a bear." Immediately, his master's wife cast her eyes upon Joseph. This use of the word *bear* is also found in *Numbers Rabbah* 13:5.

The Talmud offers us a good description of the bear and its habits. Commenting on the verse in Daniel 7:23 "And behold another beast, a second like to a bear," Rabbi Joseph wrote that "another beast" refers to the Persians who eat and drink greedily like the bear, are fleshly like the bear, have shaggy hair like the bear, and are restless like the bear (*Avodah Zarah* 2b).

Lewysohn writes, "The Persians are compared to the bear which bolts its food, is covered with a girdle of fat and can stand the winter with but little food."[2] The skin of the bear is wooly and thick and only gets softer with age. He is always rolling about, even if kept in a cage.

One of the legends connected with Rabbi Chanina ben Dosa is that he kept goats. On being told that they were doing damage to crops, he exclaimed: "If they indeed do damage may bears devour them, but if not, may each of them at evening time bring home a bear on their horns." In the evening each of them brought home a bear on their horns.

A story was told of two men sunning at the edge of a big forest. One had a big towering frame, and the other was small. Fascinated by the bulk of the other man, this small man said, "What a man you are. Do you know what I would do if I was as big and as strong as you? I'd go into the forest and find the biggest bear and tackle him alone." The big fellow looked down on the small man and said, "There are plenty of small beasts in the forest."

The story is told that Rabbi Uri Strelisker in company with several *chasidim* traveled by coach through a wood. Suddenly the horses halted, and the driver saw a bear coming toward them. The rabbi walked up to the bear and gazed steadily into his eyes. Thereupon, the bear turned and walked away. The rabbi said that it was not a miracle for the Torah explicitly states: "And the fear of you and the dread of you shall be upon every beast of the earth" (Genesis 9:2). "No unspoiled man need fear any beast," he added.

Leib Spoller, "The Grandfather," told the following experience of his youth. He once arrived in a small town and heard that the nearby Polish count had jailed a Jew for incurring a debt. He also discovered that it was the custom of the count to compel the victim to dance at the ball he held on his birthday. If the prisoner danced satisfactorily, he would be freed. The rabbi was determined to free the Jew and secured a dancing instructor, and soon he became proficient in the art of dancing. On the night of the ball the rabbi crept into the dungeon, and exchanged clothes with the prisoner. When the guards arrived to take the prisoner to the ball, the rabbi was unnoticed because of the dim lighting. He was then ordered to wear a dried bearskin and to dance opposite the village overseer. He was told that if he failed to perform the correct steps he would be beaten. He danced perfectly and so enabled the release of the prisoner. This is the origin of the Yiddish proverb, "One can even teach a bear to dance."

Proverbial Sayings

Where there is no forest there can be no bears (*Sotah* 47a).

When a person enters a town and is accosted by a tax collector, it is as though he had met a bear (*Sanhedrin* 98b).

As a roaring lion and a ravenous bear, so is a wicked ruler over a people (Proverbs 28:15).

Bee

The bee in Hebrew, "*deborah*," was a popular female name in the days of the patriarchs; thus the nurse of Rebekah was named Deborah (Genesis 35:8). Famous in Jewish history was Deborah, one of the seven prophetesses whom God raised in Israel; she judged Israel under the palm tree of Deborah between Ramah and Beth-El (Judges 4:4).

The word *deborah* is probably derived from the root *daber*, "to speak," and may refer to the constant humming noise of the bee.

The bee is mentioned often in the *Tanach* but it is questionable whether bee cultivation originated with the Jewish people. In Deuteronomy 1:44 the wild bee is compared to a hostile army; this is understandable when we realize that the average swarm of bees numbers about 30,000; compare Psalm 118:12: "They compass me about like bees"–surrounded on all sides by large numbers of bees.

In Isaiah 7:18 there is a possible reference to enticing the bee into a hive. An overt reference to bees is found in Judges 14:8, where a swarm of bees is found in the body of a dead lion. Normally bees would not approach a dead carcass for they shun anything that has a bad odor, but in this instance the sweltering heat of the summer dried up the body of the lion and attracted the bees to nest there. Herodotus records how bees and honeycombs were found in a skull.

As far as organizational skill and ability are concerned, bees have been compared to ants, and it is interesting to remember that whereas ants are called a people (Proverbs 30:25), bees are designated an *edah*, a congregation. The B. D. B. Hebrew Lexicon defines *edah* as "properly a company assembled together or acting concertedly." This seems to describe the work of the bees. It is worthy of note that the Septuagint on Proverbs 6:8 supplies us with the added information that the qualities ascribed to the bees are also found in the ants. Both people and congregation are honorable titles. The bees are not only called a congregation, but they also serve the congregation or community, for they have been responsible for sweetening the palates of the world community with their honey. Until the eighteenth century, honey was the basic source for sweetening.

Canaan is described as a land "flowing with milk and honey," a biblical expression often found in *Tanach* that suggests that large quantities of honey were collected from wild bees making their homes in rocks and hollow trees. In Psalm 81:17 it is written, "And with honey out of the rock would I satisfy thee." Milk and honey are products of a land rich in grass and flowers. It seems that both articles were abundantly produced in Canaan even when it was in a state of devastation (see Judges 14:8 and 1 Samuel 14:26).

The Hebrew word for honey, *devash*, is found no less than forty-eight times in *Tanach*; sometimes the word denotes fertility and abundance. It is also used in a figurative sense as in Psalm 19:11 where the ordinances of the Lord are described as "sweeter than honey and the honeycomb"; in Genesis 43:11 and Ezekiel 27:17 there is a reference to honey being exported.

In the talmudic period apiculture was a recognized industry, and honey was considered a very precious commodity; it was a sixtieth as sweet as the biblical manna (*Berachot* 57b), and to children manna had the taste of honey (*Yoma* 75b). A large number of preparations included honey, and there was both a beverage and a food consisting of honey. The former was called *inomelin* (*Shabbat* 139b) and the latter *rihata* (*Berachot* 37b). Honey was also used for medicinal purposes (*Berachot* 44b, *Bava Metzia* 38a). Many people today follow this talmudic example for health reasons. The Talmud explains how to render bees impotent by giving them mustard leaves (*Bava Batra* 80a), and according to one writer "this proves that the Jews in Babylonia were skilled in apiculture."[1]

However, honey was strictly forbidden to be used in the sacrificial rites as formulated in Leviticus 2:11: "You shall make no leaven nor any honey smoke as an offering made by fire unto the Lord." Commenting on this verse Maimonides observes that idolaters chose sweet things for their sacrifices, which they seasoned with honey. Our law therefore forbade us to offer leaven or honey.[2] This view of Maimonides is corroborated by the following: the Incas of Peru offered up honey as a sacrifice to the sun, the Babylonians built their temples on ground consecrated by honey, and Kama, the Hindu god of love, used a bow, the strings of which were made of bees.

In Jewish tradition honey is used on the eve of Rosh Hashanah, the New Year, when we dip apples in honey to symbolize the sweetness of the approaching year.

A very beautiful and fascinating custom was followed in the twelfth and thirteenth centuries by several communities in France, Germany, and England. On Shavuot, which commemorates the giving of the Torah on Sinai, the Jewish child at the age of three was escorted to the synagogue to receive his first Hebrew lesson. He was placed in a position of honor on the *bimah*, and at the conclusion of the service, a slate (with some of the Hebrew letters of the alphabet written on it the previous day) was brought to the child. Each letter was smeared over with honey, and as soon as the child repeated after the rabbi the name of the letter, he was allowed to lick the honey covering the letter. What an unforgettable lesson! Normally learning the Hebrew alphabet can be a dull exercise, but through this custom the child was introduced to the Torah, which became literally as sweet as honey.

If the child is naturally attracted to sweet things, so Jewish history has furnished us with many examples of outstanding personalities who have been inspired by Torah–true Judaism–to see design, wonder, purpose, and sweetness in God's creation. One such distinguished figure is the Jewish medieval philosopher Joseph Albo, who was enchanted by the ways and habits of the bees and made a profound and extensive study of them, as we learn from the following passage which is replete with the sweet wisdom and abounding love of Divine Providence:

> The bee would not so far as the nature of its own substance is concerned, have the intelligence to build the waxen cells containing the honey in the form of a hexagon, but it owes this intelligence to God. The advantage of the hexagonal form is that it is similar to the circle which is the natural form but this is superior to the circle because in the figure of a hexagon a body that is made up of adjacent hexagons has the vacant space between them all filled and there is no empty space left, whereas if the cells were circular in the shape of a cylinder and placed in juxtaposition there would remain a vacant space between them that would be wasted![3]

Even more descriptive and informative is the following remarkable passage:

> The bee would find the nectar useless if she could not transport it; therefore she possesses a sac in which to carry the nectar to the hive. But nectar is a light substance which evaporates and cannot be stored; therefore the bee possesses a tiny factory in her body which produces an enzyme which is injected into the nectar and causes it to congeal into honey. However, honey must be properly stored to endure and the bee needs some place to keep her young ones, so she therefore possesses another tiny laboratory which produces wax. But animals relish honey and against their depredations the beehive would not be safe; therefore the bee is equipped with a flaming sword to repel invaders and so another little laboratory produces an irritating poison which the bee injects with its sting into the body of its enemy. However, if the bee were armed with such a weapon, it would become a menace to the world. Therefore, its sting is barbed and cannot be withdrawn and the bee is thereby eviscerated; it stings but once and falls dead; only the Queen Bee, which must repeatedly use its sting to kill off the unneeded bees, possesses an unbarbed sting.[4]

Finally there is a halachic ruling: the bee as an insect is forbidden to be eaten, but the honey it produces is kosher because it originates in the flowers and the bee is no more than a synthesizing agent (compare *Bechorot* 6a, *Chullin* 63a, *Yoreh De'ah* 81:1).

Proverbial Sayings

As the bee gathers for its owner, so Israelites accumulate merits and good deeds for the glory of their Father in Heaven (*Deuteronomy Rabbah* 1).

Men say to the bee, "Neither of your honey, nor of your sting" (*Numbers Rabbah* 2).

Birds

Birds are often mentioned in *Tanach* and are designated either as *of*, fowl, or as *tzipor*, bird.

We find many references to the nesting of birds. Thus, the law forbidding the sending away of the mother bird (Deuteronomy 22:6) is referred to as *kan tzipor*, a "bird's nest." The Book of Proverbs writes, "As a bird that wanders from her nest is a man that wanders from his place" (27:8), and the Psalmist observes "wherein the birds make their nests" (Psalm 104:17).

In addition to nesting, we read of the singing of birds; compare "and one shall start up at the voice of a bird" (Ecclesiastes 12:4) and "the time of the *zamir* is come and the voice of the turtle is heard in the land" (Song of Songs 2:12). The turtle is called *tor* in Hebrew because of the sound of the note it utters, and it is a natural to connect *zamir* with *zemer*, "song." Feliks observes that the name *zamir* does not refer to a particular bird but is a generic name for several singing birds that are found in Israel.[1]

Large and small birds are differentiated; the large may attack with their claws at night, but the small sleep at night and arise with the dawn. As the Aramaic word for morning is *tzafrah*, we call the small bird *tzipor* to denote that it is a morning bird.

Halachically, birds are divided into two main groups: the clean versus the unclean or impure. The clean birds such as the dove and turtledove do not possess cruel habits; they are tame, docile, and patient. The marks of identification of clean birds are not mentioned in the Torah, but are discussed in the Talmud. These marks include a projecting claw that is longer than the others, a crop, and a stomach that has a membrane that can easily be peeled off. An additional qualification is that a bird that associates or dwells with unclean birds becomes unclean: "Not without reason does the starling go to the raven, they are of the same species" (*Chullin* 65a).

Rabbinical literature has provided us with detailed information that helps identify the unclean bird. Thus, birds that handle their food with their claws while eating the victim are forbidden (*Yoreh De'ah* 82:2). Birds that perch on a pole or rope and stretch out their two toes to each side or ones that snatch and eat pieces of food directly from the air instead of placing them on the ground are forbidden (*Yoreh De'ah* 82:2).

There are twenty-nine classes of unclean birds; the clean are without number (*Chullin* 63a).

One of the most striking aspects of bird life is migration that takes them to warmer climes. Birds will travel only when the sky is clear, and when they are overtaken by mist or cloud they will either rise above the cloud or descend and wait for better weather. It seems incredible how the migrants know their way over stretches of unknown land. Like experienced pilots these birds seem to be equipped with computerized brains that instinctively guide them to their destination.

Migration is a puzzling mystery that is beyond our finite minds. Surely the finger of God is evident in it. "How does the inexperienced migrant fly quite alone on its first migration thousands of miles across the equator, often at a confident speed of several hundred miles a day and adjust itself to the very different sky patterns in the southern hemisphere which it has never seen before?"[2]

Or consider the golden plover, which migrates annually from Alaska and Siberia to the Hawaiian Islands, a distance of 2,000 miles over open ocean with no landmarks to guide it. The young birds who have never made the flight do so unerringly and return in the summer to their original homes in Alaska and Siberia. God has given the birds a mysterious power to accomplish this feat, to fly nonstop without food or rest to reach their destination.

In the *Tanach* there is explicit reference to the migration of birds. The prophet Jeremiah writes, "The stork in the heaven knows her appointed times, and the turtle and the swallow and the crane observe the time of their coming but My people know not the ordinances of the Lord" (Jeremiah 8:7). Jeremiah marvels at the innate wisdom of the birds who, with precise knowledge, observe the time of their coming and going to different lands. And in the Book of Proverbs we read, "As the wandering sparrow, as the flying swallow, so the curse that is causeless shall come home" (Proverbs 26:2).

We know from the Bible that quails gathered in large numbers on the shores of the Mediterranean in the months of September and October to migrate to Asia and Africa where the weather was warmer and they returned with the wind (Exodus 16:13, Numbers 11:31, Psalm 105:40).

Similar to migratory birds, carrier pigeons carry letters fastened under their wings and can cover a distance of 300 miles in two hours. An interesting Midrash informs us that a homing pigeon will always return to its base, no matter in which direction it has gone (*Song of Songs Rabbah* 4:1).

Pigeon racing is mentioned in the Talmud and is treated as a form of gambling. The Rabbis frowned upon this sport, and those who indulged in it were ineligible to give evidence in a court of law (*Sanhedrin* 24b, 25a). Rashi interprets the mishnaic expression "pigeon trader" to refer to people who train pigeons to fight one another. This would probably be a form of cockfighting, which is forbidden in Jewish law.

All three types of birds–migrants, race pigeons, and carrier pigeons–seem to exhibit a superintelligence that may have given rise to the notion that birds possess souls. An echo of this idea is found in the Book of Psalms: "In the Lord have I taken refuge; how say you to my soul; flee you! To your mountain, you birds?" (Psalm 11:1). Primitive man regarded the bird that flies through the air as a symbol of the soul, and on Egyptian monuments the soul of the king is represented as a bird.[3]

Many are the customs, stories, and parables surrounding birds. One beautiful legend concerns the manna that our ancestors ate in the wilderness. When the heavenly food first appeared in the desert, Moses commanded, "Six days shall you gather it but on the seventh day is the Sabbath, in it there shall be none" (Exodus 16:26). Now a number of base and ungrateful people who rebelled against the authority of Moses endeavored to prove him to be a false prophet. They therefore took their portion of manna and spread it on the ground early on the Sabbath day. Immediately the birds swooped down from the skies and cleared the fields so that there was not a vestige of manna left.

In response to this timely and miraculous intervention of the birds, the custom arose that on *Shabbat Shirah* when the story of the manna is read in the synagogue, we remember with gratitude the prompt action of the birds and scatter gruel or any similar food outside our homes and feed the birds. In this manner we demonstrate our eternal appreciation to the birds of the field. Indeed some follow this custom throughout the year. This meaningful and pleasant custom is discussed in the *Minhagei Yeshurun*.

The Chafetz Chayyim devoted his life to writing about the dangers of *lashon hara*, the evils of slanderous talk. Referring to the leper and the ceremony of "the two living clean birds" (Leviticus 14:4), the Rabbis commented that these birds symbolize the leper's evil tongue: "As the birds chirp and chatter so did he (the leper) babble and prattle. The voice of the bird shall thus affect forgiveness for the words of calumny."

An interesting Midrash throws some light on the above ceremony. When the Pharaoh was smitten with leprosy he was advised by his counselors and magicians to slaughter not birds but innocent Jewish children every day and bathe in their blood to cure him of leprosy (*Exodus Rabbah* 1:41).

One aspect of bird life that has not yet been considered is bird-watching. This is a popular pastime today, but it was not unknown in the early history of our people. The graphic description of bird life recorded in Bible and Talmud is proof that our ancestors closely watched the ways and habits of birds. Since the reemergence of the State of Israel there are many areas in which bird-watching is indulged. Near fish ponds, lakes, and rivers are found a large variety of birds. A special treat is in store for those who visit the area north of Massada on the Dead Sea where there are birds of special varieties–red-winged glossy starlings are everywhere–and during migration periods Ein Gedi teems with birds of passage. Indeed, Israel is a small country, but many areas can claim to be a bird-watchers' paradise.

Jewish literature is replete with anecdotes and parables regarding birds. Those interested in kabbalistic doctrine should acquaint themselves with the famous parable of the birds as expounded and elaborated by Rabbi Moses Cordovero (1522–1570) in his well-known mystical work, *Tomar Devorah*. Another classic of Jewish literature is the letters and diaries of Gluckel of Hameln (1646–1724), who narrates the following story:

A mother bird once set out to cross a windy sea with her three fledglings. The sea was so wide and the wind so strong that the bird was forced to carry her young one by one in her claws. When she was halfway across the sea with her first fledgling the wind turned to a gale and she said: "My child, look how I am struggling, I am risking my life for you; when you are grown up will you do as much for me when I am old?" "Oh yes, I will do everything you ask of me," the little bird replied, "only bring me to safety." Whereupon the mother bird promptly dropped her offspring into the sea. The bird then returned to the shore and set forth with her second fledgling. The mother repeated the same question and as she received a similar reply as the first, again she dropped it into the waters below. When the mother put the same question to the third fledgling, the little bird replied, "My dear mother, it is true that you are struggling mightily and risking your life for me and it would be wrong not to repay you when you are old, but I cannot commit myself. However, I can promise that when I am grown up and have children of my own, I shall do as much for them as you have done for me." The mother was satisfied and carried her ashore safely.

Finally, we have a chasidic story:

A man once came to the Lekhivitzer Rabbi and admitted that he was aware of his imperfections; his only consolation, he said, was his knowledge that others were even inferior to him. The rabbi would not accept this and told him a parable. A king had an orchestra that regaled him with music. He also possessed a nightingale that sang at intervals. The king found himself rejoicing more in the natural untutored melodies of the little bird than in the studied harmonies of his orchestra.

Similarly the King of Kings has hosts of angels who sing before Him in perfect harmony, yet He prefers to hear the imperfect and even discordant prayer of mortal beings on earth. As long as we offer our services to the best of our ability, we need never feel disheartened at our inadequacies.

Proverbial Saying

One bird tied up is better than one hundred in the air (*Ecclesiastes Rabbah* 4:9).

Camel

The Hebrew name for camel is *gamal* and is derived from *gomel*, which in biblical Hebrew means "to repay good or evil." This reflects the nature of the animal, which is loyal and helpful to its master in many ways. Thus, the camel will kneel down to receive its rider and will return to the same position during loading and unloading, for the camel is used both for riding and carrying burdens and the kneeling of the beast was essential as the average height of the animal was about six to seven feet. Furthermore, the camel senses when a sandstorm is about to erupt and will quicken its pace in order to reach a place of refuge quickly so that his master is not caught in the storm.

Indeed, the camel is of service to mankind in a variety of ways. Its hair was used for the making of clothes as reported in *Menachot* 39b and in *Shabbat* 27a where it is alluded to as the "wool of camels" and we are reminded that it was not to be mixed with sheep's wool (*Kilayim* 9:1). The hair was also used for the manufacture of tents, saddle bags, and sandals because it was durable. In addition, sal-ammonia is produced from its urine, and its dung is used for fuel among the Bedouin.

The Talmud teaches that, at the time of mating, the camel can be ferocious and even attack another animal and kill it. "In the case of a camel which 'covers' among other camels and a dead camel was found at its side, it is obvious that the one killed the other" (*Bava Batra* 93a). In another instance we learn that "if among camels there is a lustful one and a camel is found killed by its side, it is certain that this one killed it" (*Sanhedrin* 37b).

According to the *Jewish Encyclopedia*, the *gimel*, which is the third letter of the Hebrew alphabet, is perhaps so called because the shape of the letter in the ancient west Semitic script bears a resemblance to the neck of the camel.

The camel is primarily a beast of burden and is specially suited for desert country. Though it is a large and heavy animal, it can cover 30 miles a day with a load weighing nearly a half-ton. Nature provides the camel with feet that are covered with a tough spongelike substance that makes it possible for it to traverse the burning sands of the desert. Thus, it has earned its name–"the ship of the desert."

Another distinctive feature with which the Almighty has blessed the camel is the possession of a fifth stomach, which serves as a reservoir of water. Such is the bodily construction of the camel that the water is kept pure and sweet for journeys lasting eight to ten days. When the camel is thirsty, it automatically compresses the stomach and forces the water into the upper stomach and so satisfies its thirst.

By nature, the camel is patient and gentle and is different from other animals inasmuch as it is satisfied with a poor diet: it subsists on thorns and thistles. These prickly plants do not cause it pain as nature again comes to its assistance, as its lips, tongue, and palate are covered with hard, thick skin. Furthermore, nature has provided the camel with a long neck proportionate to the length of its thighs so that it is able to pick up its food easily.[1]

In biblical times, camels were considered a valuable possession. Thus, Abraham included camels among his stock (Genesis 12:16). Of the Midianites and Amalekites, the Bible says, "Their camels were without number, as the sand which is upon the seashore for multitude" (Judges 7:12). Job was blessed in the latter end of his life with good camels (Job 42:12), and in Ezra there is a reference to 435 camels (Ezra 2:64). In war, camels were included in the spoils (1 Samuel 27:9), and the Chaldeans "fell upon the camels and have taken them away" (Job 1:17).

The Talmud relates that there are two kinds of camel, the Persian and Arabian, and they are distinguished by the thickness of the neck (*Bava Kamma* 55a). In *Ketubot* 67a, we learn that camels could be levied for a wife's *ketubah* (marriage contract).

The dromedary is known in the Talmud as the flying camel (*gamal parchah*), and Isaiah refers to the "swift beasts" (Isaiah 66:20). The flying camel figures prominently in an interesting talmudic discussion regarding two sets of witnesses who contradict each other in a criminal lawsuit. The ruling of the court revolves around the feasibility of the witnesses to travel from Sura to Nehardea, a distance of more than twenty parasangs in one day. If they traveled on flying camels, the distance could be covered in one day and their evidence could be accepted; otherwise, the witnesses were not telling the truth (*Makkot* 5a). A similar case is reported in *Yevamot* 116a. The *Encyclopaedia Britannica* affirms that "the fleeting camels will carry their rider and a bag of water for 15 miles a day without a drink."

The name "flying camel" was given to the Palestine Flying Club founded in Tel Aviv in 1933; one of the main activities of this club was gliding, and one of its aims was to turn its members into flight instructors.

Reckless and impatient drivers on the road should heed the advice offered by our Sages of old, who cautioned camel drivers and even laid down guidelines governing correct behavior on the road. Two camels attempted to ascend Beth Horon, a dangerous road that cut through rock in a zigzag fashion. "If they both ascend at the same time they will tumble down into the valley below. If they ascend after each other both can go up safely. How should they act? If one is laden and the other unladen, the latter should give way to the former. If one is nearer to its destination than the other, the former should give way to the latter. If both are equally near to or far from their destination, make a compromise between them. The one which is to go forward compensates the other which has to give way" (*Sanhedrin* 32b).

An interesting story regarding a camel is reported in the Talmud. It once happened that two Jews were taken captive on Mount Carmel and their captor was walking behind them. One of them said to the other, "The camel walking in front of us is blind in one eye and is laden with two barrels, one of wine and the other of oil, and of the two men leading it one is a Jew and the other a heathen." Their captor said to them, "You stiff-necked people, whence do you know this?" They replied, "Because the camel is eating of the herbs before it only on the side where it can see, but not on the other, where it cannot see. It is laden with two barrels, one of wine and the other of oil, because wine drips and is absorbed into the earth, while oil drips and rests on the surface. And of the two men leading it, one is a Jew and the other a heathen, because a heathen obeys the call of nature in the roadway, while a Jew turns aside." He danced before them and exclaimed, "Blessed be He who made choice of Abraham's seed and imparted to them of His wisdom." He then liberated them (*Sanhedrin* 104a–b).

The Sassover Rabbi expounded on a statement in the Talmud as follows: The Rabbis knew not the meaning of the verse "Cast your burden upon the Lord and He will sustain you" (Psalm 55:23) until a traveling merchant told them (*Megillah* 18a). The rabbi explained: the sages did not know whether the verse permits a man in need of a helping hand to combine his trust in the Lord with a request for help from the people. Without being asked for the favor, the merchant said, "Throw your burden on my camel," at the very moment they were discussing the verse. They received it as an omen that the verse implies that the man should not ask for the favors of mortals but should petition the Almighty alone.

The *Mechilta* relates that Rabbi Tarfon and the elders were sitting in the shade of a dove-house in Javneh and discussed the verse in Genesis 37:25 "with their camels bearing spices and balm and laudanum." They suggested that this verse makes known how very much the merit of the righteous is of help to them. For would they not have killed this beloved friend (Joseph) of Arabs with the smell of their camels and the smell of their itiron, but God arranged it for him that there be sacks full of spices of the itiron (*Mechilta*; compare Rashi on Genesis 37:35).

It is interesting that a camel was involved in choosing the burial place for Maimonides. He lived and died in Egypt, but requested that his last remains be interred in the Holy Land. When the time for burial arrived, his body was placed on a camel and transported to the Land of Israel, and when it reached Tiberias, the animal stubbornly refused to go further. The authorities therefore had no alternative but to bury Maimonides in a plot of ground chosen by the camel. However, the grave was not in a deserted or isolated ground, but close to the sepulcher of Rabbi Johanan ben Zakkai.

Proverbial Sayings

In Media a camel can dance on a bushel basket—meaning that everything is possible (*Yevamot* 45a).

Many an old camel is laden with the hides of younger ones (the old may survive the young) (*Sanhedrin* 52a).

One does not say, examine the camel only examine the lamb (*Chagigah* 9b).

As the camel, so is the burden (*Sotah* 13b). This saying can be interpreted in a variety of ways: the more eminent a person, the greater is his responsibility. In another context the simile of the camel is used in regard to charity. A rich man who does not dispense charity commensurate with his wealth, of him it is said, "in accordance with the camel is the burden"—the richer the person, the more is expected of him.

Cat

The cat is not specifically mentioned in the Bible because the domesticated cat came upon the scene rather late. However, the cat was worshiped in Egypt and domesticated in Greece and Rome before the Common Era.

The cat is found in the Talmud where it is known either as *chatul*, a general name, or *shunra*. The name *chatul* is probably derived from the root *chatal* found in the Bible twice, once as a verb in Ezekiel 16:4 meaning "to enwrap or swaddle" and once as a noun *chatulah*, a "swaddling band," in Job 38:9. In this context, Shoshan designates the cat as "the wrapper" because of its habit of curving its body when it relaxes or sleeps.[1]

The designation of *shunra* is dubious, but a clue to its meaning is found in the Talmud: he who dreams of a cat in a place where the word for it is *shunara*, a beautiful song (*Shirah Naah*) will be composed in his honor; but where the word for it is *shinra*, a change for the worse (*shinnui ra*) is in store for him (*Berachot* 56b). The Munich manuscript of the Talmud has a variant reading—"Where the word for it is *surama*, a beautiful song will be composed in his honor."

In "*Chad Gadya*," the popular ballad sung on the Seder night of Pesach, the cat is the first animal mentioned: "Then came the cat and ate the kid." All commentators describe the nursery rhyme as an allegory in which the animals are employed figuratively to represent individuals or nations. Thus, it has been suggested that the kid is Joseph, who was seventeen years old when he was sold; the numerical value of the word *g'di* (kid) is 17. By nature the cat is envious, and the brothers represented by *shunra* were envious of Joseph and sold him to the Ishmaelites. Others read into *shunra* the name *Sonei Ra*, the wicked enemy, and interpret the cat to refer to Nebuchadnezzar, who destroyed the Temple in 586 B.C.E.

In one respect the cat has never changed – it always has been and still is an implacable enemy of the mouse upon which it pounces relentlessly. An interesting story is recorded in the Talmud about a cat and mice. A man borrowed a cat from his neighbor. The mice then formed a united party and killed it. The Talmud then pokes fun at the cat for allowing itself to be devoured by mice. A variant of the story is that the cat died from eating too many mice. Because the question arose whether the borrower was liable to pay for the cat, it would seem that cats had a commercial value. This is corroborated by another case in which we learn that one who is in charge of a cat belonging to another person must not give the cat water that is left uncovered for fear that the water was drunk by a snake; if the cat drank the water, it would affect the sale of the cat, which would lose some of its value (*Avodah Zarah* 30b). Although the cat is to an extent immune from poisoning it would nonetheless affect the sale price of the animal.

In addition to eating mice, cats also eat snakes. "Rabbi Papa said: 'A man should not enter a house where there is no cat in the dark.' What is the reason? Lest a snake might wind itself about him without his knowing and he would come to danger" (*Pesachim* 112b). From this we learn that a man should not enter a house in which there is a cat without wearing shoes because the cat may kill a snake and eat it. The snake has a number of small bones and if one penetrates a person's foot it would not be easy to extract and would endanger the life of the person (*Pesachim* 112b).

Writing on the habits of cats the Talmud mentions an incident when a cat bit off the hand of an infant, whereupon Rav declared that it was permissible to kill the cat and that it was a sin to keep it (*Bava Kamma*

80b). This ruling is understandable, but this was an unusual occurrence and did not contradict a statement that permitted the breeding of cats because they help keep the house clean (*Bava Kamma* 15b).

The Talmud details other interesting characteristics of cats. A cat always finds its way to its home (*Shabbat* 51b). A cat bears its young in fifty-two days (*Bechorot* 8a); medical science says fifty-five days.

The cat forgets its master on the principle that he who eats something of that from which a mouse has eaten loses his memory (*Horayot* 13a). This statement is followed by the question, Why do all cats persecute mice? Because they gnaw even at clothes; they derive no benefit because it is not food, but it entails a loss to the owner (*Horayot* 13a).

The cleanliness of the cat is obvious to all as it persistently cleanses its fur by washing itself with its tongue. However, the cat is also modest. It performs its bodily functions secretly and will consistently attempt to cover its motions. The modesty of the cat is affirmed by the famous dictum in *Eruvin* 100b; below is a different version that has been handed down. The Creator turned to the animals and said, "Cooperate with me in forming a higher being, to which all of you will contribute a desirable characteristic. The cat will contribute modesty, the ant honesty, the tiger courage, the lion bravery, the eagle diligence, and so on. Thus will man not only be akin to you but will also represent the finest in you."

The following data about the cat is informative:

Its hair stands rooted in a tiny well of oil and its scales are symmetrically arranged, tapering toward the top, a veritable feat of construction and purposeful planning. It is of flexible material, it keeps out the cold and retains the warmth of the body. It is self-oiling and can be renewed from its roots. It is water resistant and shields the skin against blows and abrasions and in many cases supplies protective coloration. The cat has eyes specially constructed for night vision, and its whiskers enhance the function of smelling. It possesses sharp daggers of tough horn which can be retracted and kept out of the way when not in use, else they would make the animal's footsteps heard by its prey. This is obviously constructed for the purpose of destroying mice.[2]

During the dark Middle Ages when the Jew was feared as the devil, the cat played a special role in sorcery, divination, and witchcraft. In some German provinces, a toad or a cat was believed to be found ensconced on the "altar" of synagogues, thereby according the devil, whom these creatures represented, a position analogous to that of Christ in the church;[3] this belief was held until quite recently. Jews were also alleged to practice magic, and nursery rhymes have preserved the tradition that Jews could and often did turn themselves into cats.

Finally, the ineffectiveness of Jewish conversions to Christianity is known by the following story:

> A converted Jew in Cologne, who because of his apparent piety and devotion, was eventually appointed Dean of the cathedral. Yet after his death when his will was opened, it was found that he had ordered the erection of the figures of a cat and mouse on his grave to indicate that a Jew can as little become a Christian as the two animals can live together on friendly terms. The same thought is expressed in the Freising cathedral where there is a picture of the judensau with the inscription, "As surely as the mouse never eats the cat, so surely can the Jew never a true Christian become."[4]

Proverbial Sayings

A sleeping cat does not catch the rat (Ibn Ezra).

If cats had gloves, they would catch no mice (Ausubel).

Cock and Hen

The cock is known by one of three names–*sekhvi*, *gever*, and *tarnegol*.

The word "*sekhvi*" is mentioned once in *Tanach*: "Who has put wisdom in the inward parts?" or "Who has given understanding to the *sekhvi*?" (Job 38:36). Rashi states that *sekhvi* is a *tarnegol*, "a cock." Some translate the word as mind or heart, but tradition seems to favor "cock."

Today, we are familiar with the word *sekhvi* because a special blessing has been allotted to it among the early morning blessings in our prayer book. Indeed, it heads the list of benedictions, for we thank God for giving the *sekhvi* understanding to distinguish between day and night. The time of day and night changes with the different continents, but the Almighty has implanted in the cock the instinct to crow at dawn wherever it may be. It is worth noting that in the days of the Temple the service began with the cleansing of the altar at the cockcrow (*Yoma* 20b). Today prayer is substituted for sacrifice and the altar, and it is a salutary thought that we begin our prayers with the meaningful blessing that symbolizes the cleansing of our mind and heart.

Some suggest that the inclusion of this blessing in the liturgy is due to Persian influence, but Israel Abrahams, a modern scholar, refutes this and states that the lesson underlined here is the regular recurrence of daily phenomena. On awaking, the worshiper expresses his sense of the order of nature and the marvelous regularity of her operations. The Talmud translates Job 38:36 as follows: "Who has put in the inward parts (of man) or who has given understanding to the cock" (*Rosh Hashanah* 26a). Thus, both the intelligence of man and the instinct of the whole animal world are derived from the same divine source.[1]

In a striking passage of the *Zohar* we learn that the cock crows at midnight and early morning. In poetic imagery the *Zohar* depicts for us the heavenly scene at midnight when the divine flame strikes against the wings of the cock who then crows and God enters the Garden of Eden and has joyous communion with the souls of the righteous. Again at daybreak when the sun appears, Israel takes up the song below in unison with the sun above. Rabbi Eleazar said: "Were mankind not so obtuse and insensitive, they would be thrilled to ecstasy by the exquisite melodiousness of the orb of the sun when he journeys forth singing praises to God" (*Vayakhel* 196a).

The above is a curtailed version of the original and calls for clarification. The *Zohar* reminds us indirectly that it is proper to recite at midnight *Tikkun Hatzot*, the anthology of prayer and study reserved for the pious. More pertinent to our subject is the reference to the break of dawn and the appearance of the sun. Then in the stillness of the morning when the air is pure and clean and the wheels of industry have not yet begun to move, the crow of the cock arouses us from our slumber and calls us to prayer. In our material world we are apt to use mechanical gadgets for many of our needs and requirements. Judaism, however, reminds us very forcibly that the day should begin not with the artificial sound of the alarm clock, which is made with the hands of man, but with the crow of the cock, the heavenly bell, the handiwork of God. The Torah often reminds us that the day should be initiated with a spiritual exercise. The first action leading to the momentous *Akedah*, the binding of Isaac, was the early rising of Abraham, probably at dawn: "And Abraham rose early in the morning" (Genesis 22:3). The early morning is reserved for prayer as the Psalmist records: "O God, in the morning shalt Thou hear my voice; in the morning will I order my prayer to Thee and will look forward" (Psalm 5:4). Indeed in ancient times the *vatikim* (very pious people) read the *Shema* with the rising of the sun, which coincided with the cockcrow (*Berachot* 9b, 25b–26a), and we have in our midst today pious individuals who follow this practice.

In addition to the lessons adduced from the crow of the cock, the Talmud underlines a most vital aspect of its life–the cock is loyal, faithful, and chivalrous to his mate. In the animal world every species fends for itself and searches for food, but the cock provides for his wives. Moreover, in the words of the Rabbis the cock does not force his attention on the hen, but coaxes her. How does he coax her? He says to her: I will buy you a cloak that will reach to your feet. After the event he tells her: may the cat tear off my crest if I have any money and do not buy you one (*Eruvin* 100b). This detailed portrayal of the relationship of the cock and his wives is very illuminating, and it contains a timely lesson for mankind.

This leads us to the second name by which the cock is known: *gever*. This word is found in the mishnaic expression "*Keriat hagever*," the crowing of the cock (*Yoma* 20b). Here the cock bears the name of man. It has been suggested that the family life of the cock almost resembles that of mankind; hence, the name *gever*, which incidentally is connected with *gevurah*, strength, or might. Indeed, the crow of the cock is a unique and mighty force in the animal world, as is illustrated by this talmudic statement: If a cock stretches its head into the cavity of a glass vessel he can break it by means of his crowing (*Kiddushin* 24b). It should also be noted that the word *gever* in Isaiah 22:17 is compared by Rashi to a cock.

Tarnegol is the third name by which the cock is known. It is an Aramaic word and is widely used in the Talmud. The derivation of the word is uncertain, but the Rabbis give it a fanciful interpretation when they observe that he who sees a *tarnegol* in a dream may hope for a male child; he who sees a hen in a dream may hope for a fine garden and rejoicing (*Berachot* 57a).

Tarnegol is also traced to *Nergal*, a Babylonian deity mentioned in 2 Kings 17:30. Thus the Rabbis interpret the words: "And the men of

Babylon made Succoth-Benoth," the image of a fowl: "And the men of Cuth made Nergal," that is a cock (*Sanhedrin* 63b). From this connection may derive the rule that the Rabbis formulated forbidding the selling of a white cock because it was a recognized offering of the poor to idols. However, if the cock has its spur clipped, it may be sold because a defective animal is not sacrificed to an idol (*Avodah Zarah* 14a).

Not all of the characteristics of the cock noted by the Rabbis were favorable. Thus, we learn that a certain cock killed a child by picking at its scalp, and the cock was stoned (*Eduyot* 6:1). We are also informed that the cocks of Bet Bukya (Upper Galilee) were fierce and would not allow the intrusion of a stranger among them (*Yevamot* 84a). Again, the Rabbis testify that among the birds the cock is distinguished by its fierceness (*Betzah* 25b).

This fierceness of the cock was probably responsible for the introduction of cockfighting, a sport indulged in by some people but forbidden in Jewish life. Incidentally, the fighting of a bird called *zarzir-motnayim* (Proverbs 30:31) is mentioned in *Yalkut Shimoni* (Proverbs 963). This bird is usually translated as "greyhound," but the Targum and Septuagint render it as "cock."

In the Middle Ages the cock for a male and a hen for the female were used in *Kapparah* (atonement), a rite that took place before Yom Kippur. Today we also use money in place of the animals in performing this rite.

Finally it is worth noting that *Tarnegolah* is a name of a place or district. Thus, we have Fort Tarnegola and Tarnegola of Caesarea.

Tarnegolet (the Aramaic word is *tarnagolta*) is the usual name for a hen. However, the Talmud also mentions *gabrit*, a denominative of *gever* (*Shabbat* 67b). The hen (chicken) is not specifically mentioned in *Ṭanach*, but as noted above, the Rabbis designate "Succoth-Benoth" in 2 Kings 17:30 as hens with chicks. In another biblical reference the expression "*barburim avusim*" (1 Kings 5:3) is translated as fattened fowl or chickens. Both Talmud and Rashi explain this term to mean chickens or birds fattened by force (*Bava Metzia* 86b). The engraving of hens is found on ancient Babylonian pillars, and Caesar found chickens in Britain in the middle of the last century before the Common Era.[2]

Among domestic birds, chicken and geese are the most common and are mentioned often in the Talmud. Thus we learn that chickens and geese may be caught on a festival day because it is not considered "capture" and so is not prohibited, whereas other birds are forbidden (*Betzah* 24a). Chickens were often reared because of their food value, and one Rabbi Amemar considered them to be the finest bird (*Bava Metzia* 86b). The popularity of the chicken can be gauged by the fact that the egg always referred to in the Talmud is the egg of the hen (*Shabbat* 80b).

The perennial question as to which came first, the chicken or the egg, was simply solved by the talmudic statement that all works of creation were brought into being in full-grown stature, in complete understanding, and in their designated shape and form (*Rosh Hashanah* 11a).

It is recorded that Rabbi Simeon ben Halafta, who was a keen observer of animal life, experimented not only on ants but also on hens. For instance he cured a hen that suffered from a dislocated hip bone by attaching a reed to it; on another occasion he experimented with a hen that had lost its feathers and new feathers grew (*Chullin* 57b).

A delightful story is told in the Midrash in which we learn how an invited guest who joined a family meal was asked to carve a chicken and apportion it to the members of the family, which included, in addition to the host and hostess, two sons and two daughters. He gave the head of the chicken to the host who was the head of the family, the entrails to the hostess because children issue from the womb, the two thighs to the two sons who are the pillars of the home, the two wings to the two daughters who in the future will fly away and go to their husbands, and the guest took for himself the body shaped like a boat as he came, and will leave, in a boat. This pleased the host who tested the guest and found him not wanting (*Lamentations Rabbah* 1:4).

An interesting episode about kindness to animals is recounted by Rabbi Isaac Luria (1534–1572). After having enjoyed the hospitality of his host, he asked if he could repay him in some way. The host pleaded that he was childless and prayed for an offspring. Rabbi Luria then intimated that his lack of children was due to his cruelty to animals.

The wife of his host had inadvertently removed a ladder leading to a cistern, thus depriving the chickens of water. When this was rectified, the child arrived in due course.[3]

The Besht once said: "A farmer held an egg in his hand and mused, I shall place this egg under a hen, I shall raise up chickens and shall hatch other chickens, I will sell them and purchase a cow and . . . While planning in his imagination, he squeezed the egg and it broke in his fingers. In the same manner," said the Besht, "some people are satisfied with the holiness and knowledge they have attained and constantly think they are superior to others. They do not perceive that by doing this they lose even the little they have attained."

Dog

The derivation of the Hebrew word *kelev* (dog) is unknown; the grammarian Gesenius says it is onomatopoeic, and Jastrow in his talmudic dictionary suggests that *kelev* is connected with *kalav*, meaning to make stitches resembling dog bites.

There are many types of dogs and one writer enumerates no less than twenty-four principal varieties.

The Bible deals mainly with two kinds, the shepherd dog and the wild dog. There are numerous references to flocks and herds in the Bible, and the shepherd dog must have been very helpful in guarding and protecting the flock from being attacked. Thus Job declares, "Whose fathers I disdained to set with the dogs of my flock" (Job 30:1).

Indeed the animal has earned the title watchdog, which in the English idiom can also apply to a communal leader. The lack of watchfulness is alluded to by Isaiah who speaks of "dumb dogs that cannot bark" (Isaiah 56:10).

However, in biblical days the dog had a bad reputation; he was undomesticated, wild, and ferocious. The dogs were street scavengers; they were greedy (Isaiah 56:11), and they even ate the flesh of human beings: "You shall not eat any flesh that is torn of beasts in the field, you shall cast it to the dogs" (Exodus 22:30; compare 1 Kings 14:11, 21:23). The dog is also used as a term of reproach: "What is your servant who is but a dog" (2 Kings 8:13).

When we turn to the Talmud and Midrash we find the same pattern of uncomplimentary and complimentary remarks regarding the dog. Because the preservation of life was always emphasized, we are warned not to handle a wild dog (*Genesis Rabbah* 77) nor to tolerate a mad dog, which was permitted to be killed even on the Sabbath (*Shabbat* 121b). We are also advised not to raise a bad dog in the house, as this would be a transgression of the verse in Deuteronomy 22:8, "You shall not bring blood upon your house" (*Bava Kamma* 15b, 46a).

The Rabbis even diagnosed the symptoms of rabies, canine madness: "Its mouth open, its saliva dripping, its ears flap, its tail is hanging between its thighs and it walks on the edge of the road. Some say it barks without its voice being heard" (*Yoma* 83b).

Discussing the name of Merodakh Baladan (2 Kings 20:12, Isaiah 39:1), the Rabbis assert that he was so called because Baladan was a king whose face turned into that of a dog (*Sanhedrin* 96a). Louis Ginzberg suggests that on Assyrian Babylonian monuments are found dogs in the company of Merodakh, which probably explains the expression "dog."[1]

Consider this talmudic statement: In the generation when the son of David comes, people will be dog-faced, that is, brazen and without shame (*Sanhedrin* 97a).

On the other hand Solomon declared: "A living dog is better than a dead lion" (Ecclesiastes 9:4). The exposition of this verse is discussed in the Talmud. When King David died on the Sabbath, his son Solomon sent a message to the *Bet HaMidrash*: "My father is dead and lying in the sun, and the dogs of my father's house are hungry, what shall I do?" The reply came promptly: "Cut up a carcass and place it before the dogs, and as for your father . . . carry him away" (*Shabbat* 30b).

From this interpretation we infer that it was permitted to cut up a carcass on the Sabbath in order to feed the dogs, but it was forbidden to handle and bury the dead king. In other words, the living dogs enjoy a greater preferential treatment than the dead lion, the noble King David.

The classic example of dogs being rewarded by God is recorded in the Torah: "but against any of the children of Israel shall not a dog whet his tongue" (Exodus 11:7). Instead of barking or howling and so frightening and intimidating the Children of Israel in their exodus from Egypt, the dogs remained quiet and acted as silent sentinels. For their resolute action the dogs were rewarded, and the Almighty directed that *treifah* food unfit for human consumption should be cast to the dogs (Exodus 22:30). In the words of the *Mechilta* the Lord does not withhold the reward of any creature. Moreover, the Rabbis advise that one who sees a dog in his dream should rise early and recite the verse "but against any of the children of Israel shall not a dog whet his tongue" before another verse (an unfavorable one) occurs to him (*Berachot* 56b).

Today the dog is famed for his loyalty to his master, but there is nothing new in this notion. Already in the Apocryphal Book of Tobit we find a striking instance of Jewish affection for the dog as it follows Tobias on his long and adventurous journey from home and returns with him and the angel (Tobit 6:1).

In rabbinic literature there is ample evidence testifying to the friendship and faithfulness of the dog toward its master. A fascinating story is told in the Talmud in which we learn that a number of shepherds were preparing curdled milk for a meal when suddenly they were called away. In their absence a serpent tasted the milk and in the process injected some poison into the milk. The dog who was standing by realized what had happened, and he began barking incessantly. When the shepherds returned they ignored the barking of the dog and were about to partake of the milk when the animal quickly fell upon the food, ate it, and promptly dropped dead. The shepherds were so grateful to their dog that they erected a monument over its grave and called it *nafsha d'kalba*, the dog's monument (*Yerushalmi*, *Terumot* 8, 46a).

In another passage, the *Pirke d'R. Eliezer* expressly states that the same dog that protected the flock of Abel in his lifetime also guarded the body of his master when he died. According to Jewish law and custom the human corpse must not be left unattended and is cared for by a "watcher" or watchman. From the above Midrash it would seem that the first watcher was not a human being but an animal, the dog.

The *Pesikta d'R. Kahana* records a strange incident. Rabbi Abbahu went to Caesarea to visit a friend. When the rabbi was seated, his host placed the dog next to him. The rabbi was puzzled and asked his host why he humiliated him. The host replied, "My master, I owe much to my dog, for when bandits entered the city, one of them attempted to rape my wife but the dog saved her by springing upon the bandit and biting at his testicles" (*Pesikta d'R. Kahana* 11:1).

Finally a word from one of our moralists. The story is told that a pious person once passed by a carcass of a dog. His disciples said to him, "How dreadful does this carcass smell." The master replied, "How white are its teeth." The pupils then regretted the disparaging remark they had made concerning the dog. If it is reprehensible to make a disparaging remark about a dead dog, how much more so concerning a living human being, and if it is proper to praise a dog's carcass for the whiteness of its teeth, how much more so is it a duty not to speak evil of anyone but to speak good so that it becomes a natural habit.[2]

A heathen once asked Rabbi Gamliel why God was angry only with idol worshipers and not with the idols themselves. Rabbi Gamliel answered with a parable. Suppose a king had a son who kept a dog that he named after his royal father, and suppose when his son wanted to swear, he swore by the life of his dog, thereby naming his father. Against whom would the king be angry–his son or the dog?

Proverbial Sayings

As a dog that returneth to its vomit, so is a fool that repeateth his folly (Proverbs 26:11).

Two dogs in a kennel snarl at each other, but when a wolf comes along they become allies (*Sanhedrin* 105a).

He who was bitten by a dog will tremble at its bark (*Zohar*, Exodus 45a).

A dog can't be a butcher, nor a butcher a matchmaker (Yiddish saying).

The worst dog gets the best bone (I. L. Peretz).

Show a dog a finger and he wants the whole hand (Yiddish saying).

Dove, Turtledove, and Pigeon

These three birds are grouped together because they belong to the same genus. Both the dove and the pigeon are called *yonah*, but the latter is usually introduced by the word *ben* or *bnei*, signifying either singular or plural. In Genesis 15:9 *gozal* is translated as "young pigeon," but this word can equally apply to the young of any bird.

It is conjectured that the word *yonah* is derived either from *yanah*, "to oppress or maltreat," or from *anah*, "to mourn." Both derivations are alluded to in the Bible and Talmud.

Isaiah declares, "I do moan as a dove" (Isaiah 38:14). Ezekiel refers to "the doves of the valleys, all of them moaning" (Ezekiel 7:16), and the prophet Nachum recalls that "her handmaids moan as with the voice of doves" (Nachum 2:8). In the Talmud we read that a man should always strive to be rather of the persecuted than of the persecutors, as there is none among the birds more persecuted and oppressed than doves and pigeons, and yet Scripture made them, alone of birds, eligible for the altar (*Bava Kamma* 93a). That the gentle dove was exposed to the wild beasts is underlined by the Psalmist: "O deliver not the soul of the turtledove to the wild beast" (Psalms 74:19).

The turtledove is called *tor*, probably onomatopoeic, and is symbolic of the spring: "for lo the winter is past, the rain is over and gone, the flowers appear on the earth, the time of spring has come, and the voice of the turtledove is heard in the land" (Song of Songs 2:11). The voice referred to in the above verse is possibly the cooing of the turtledove, which sounded like *tor*, hence its name.

Another characteristic of the dove–its strong flight–is highlighted in this verse: "Oh that I had wings like a dove!" (Psalm 55:7). The Song of Songs emphasizes the purity and uniqueness of the dove: "My dove, my undefiled is but one, she is the only one of her mother" (Song of Songs 6:9). Perhaps the most delightful description of the dove is found in the following passage: "O my dove . . . let me see thy countenance, let me hear thy voice; for sweet is thy voice and thy countenance is comely" (Song of Songs 2:14).

The classical text in which the dove figures most prominently is Genesis, chapter 8, where we learn that Noah sent a dove from the ark "to see if the waters were abated from off the face of the ground." Why a dove? It should be noted that the dove, more than any other bird in the Bible, is employed metaphorically in Jewish teaching, and some scholars are of the opinion that here the clean, gentle, and peace-loving dove is sharply contrasted with the unclean, cunning bird of prey, the raven. However, the dove does recognize its resting place, suggesting that Noah instinctively recognized the dove to be a reliable and trustworthy messenger. This trait is substantiated by history, for throughout the ages the carrier pigeon has proved to be a worthy and excellent courier, dispatching messages over long distances with amazing precision.

However, there is an important distinction between a dove and a carrier pigeon. The latter bird is aided by human effort, a written message that is attached to its feet, whereas the dove is not dependent upon human assistance. The dove is master of its destiny and uses initiative. In the story in Genesis it brings back a message to Noah in the form of "an olive leaf freshly plucked in her mouth" (Genesis 8:11). This expression is pregnant with meaning and interpretation. Why are we expressly informed that the leaf was freshly plucked (or torn) in its mouth? The dove could simply have picked up a leaf floating on the water, but such information would prove to be useless to Noah. However, a fresh green olive leaf freshly torn from a tree rooted in the ground indicated to Noah that the waters had considerably decreased as the olive tree is not tall. Furthermore, the olive tree is mentioned specifically because its leaves do not drop either in the days of summer or in the rainy season; the leaves are always fresh and sturdy (*Menachot* 53b). One further difficulty in this verse needs clarification. We know that the dove is particularly fond of sweet foods. Why then did it favor the leaf of an olive, which is bitter? Here, too, the dove imparts a telling message. In the words of the Rabbis the dove symbolically exclaimed, "Let my food be as bitter as an olive leaf, provided that it comes from the hands of God rather than be as sweet as honey and I be dependent upon the gifts of man" (*Eruvin* 18b). This noble sentiment is echoed in the fourth paragraph of the Grace After Meals: "We beseech Thee O Lord our God, let us not be in need either of the gifts of mortals or of their loans, but only of Thy helping hand."

In rabbinic literature the dove is invested with a host of virtues, all of which are symbolic of Israel. In our Midrash we derive a significant lesson for Diaspora Jewry. All birds apart from the dove fly, and when they grow weary they rest on the top of a tree or a rock, but the dove merely folds one of her wings and flies with the other. This characteristic has been interpreted to mean that the nations of the world need both wings, land and culture, to maintain their existence, but Israel can fly and soar to great heights even on one wing, namely, its culture, which is the Torah. In the Diaspora where she was denied her national home she was sustained by the immortal culture enshrined in the Torah (*Genesis Rabbah* 39). In another passage the rabbis declare: As the dove when it is slaughtered does not struggle, so the Israelites do not struggle when they are slaughtered for the sanctification of the Name, and as the dove saves herself only by her wings so the Israelites are saved only by the merit of the Torah (*Midrash Tehillim* 159a).

In yet another passage we read: As the dove is chaste so the Israelites are chaste . . . as the dove atones for sins, so the Israelites atone for the nations, for the seventy oxen which they offer on the Festivals represent the seventy nations so that the world may be depopulated of them . . . as the dove from the hour she recognizes her mate does not change him so the Israelites from the time they recognized the Holy One have not changed Him (*Song of Songs Rabbah* 1:63). This passage begins and ends with conjugal fidelity. This may throw some light on the nature of the sin-offering that a woman after childbirth must bring to the altar. "And when the days of her purification are fulfilled for a son or for a daughter she shall bring . . . a young pigeon or a turtledove for a sin-offering" (Leviticus 12:6). This offering indicates that the woman was pure before marriage and that she is always faithful to her partner. In this manner we equate the purity of married life among humans with the fidelity of the dove toward its mate.

An additional reason for the popularity of the pigeon and turtledove as an offering was a practical one: these birds were economical and suited the poor, this reflected in a touching episode in the life of King Agrippa recorded in the Midrash. The King decided to set aside one day to offer a thousand birds. He consequently sent a personal message to the High Priest that he alone was to offer the birds on one particular day. However, on this day a poor man appeared with two turtledoves in his hand and pleaded, "My master, the High Priest, everyday I catch four turtledoves, two of which I offer up and the other two are my livelihood. If you refuse to offer the two, my livelihood will be affected." The High Priest was moved and offered the two birds. We are informed that Agrippa received a *bat kol* (a heavenly voice) in a dream saying, "The sacrifice of the poor man rightly had priority" (*Leviticus Rabbah* 3:5).

There was, however, a period when the price of turtledoves and pigeons rose considerably, and we owe it to the foresight and courage of Rabban Simeon ben Gamaliel who would not rest at night till the price dropped. After issuing a decree, "the price of a pair of birds fell to a quarter of a demar" (*Bava Batra* 166a–b).

The price of doves was also regulated by supply; doves breed every other month, and there was a superabundance of them. This created a problem. In Babylonia and Palestine the dovecotes around the town were so numerous that fowlers had to be prohibited from snaring the birds within a distance of thirty ris from the town (about 3.8 kilometers). Indeed, laws were enacted prohibiting the setting up of cotes within a distance of fifty cubits from the town because the birds were considered a liability as they ate the seeds in the gardens and orchards. In this connection one sage explained that the amount of food a pigeon will find in a space of fifty cubits is normally enough to satisfy its hunger (*Bava Batra* 23a).

There was a variety of species of pigeons, and Jews understood the nature of these birds and could distinguish each breed. We shall mention by name two species, one the Hardisian, a domesticated dove that received its name, according to Jastrow, from the manner of its fructification (*Chullin* 139b), and the other is the Herodian dove, a domesticated indoor dove that presumably was named after Herod, who kept them in the garden surrounding his palace.

In conclusion, a story is narrated in the pages of the Talmud concerning a teacher called Elisha who was called "the man of wings." The Roman government issued a decree that any Jew found wearing *tefillin* would have his brains pierced through. Elisha disregarded the decree, put on his *tefillin*, and walked into the street. When a quaestor saw him, he attempted to flee and took off his *tefillin* and held them in his hand. Elisha was overtaken, and when asked what he held in his hand he replied, "The wings of a dove." The Rabbis ask, why the wings of a dove? Because the Congregation of Israel is likened to a dove, "*Knesset Yisrael domeh l'yonah*," as a dove is protected by its wings so is Israel protected by the mitzvot (*Shabbat* 49a, 130a).

Proverbial Saying

A people distinguished with divine precepts and meritorious deeds is compared to a dove. (*Lamentations Rabbah*)

Eagle

The common name by which the eagle is known in the Bible, where it is often mentioned, is *nesher*. However, it is generally conceded that this is a generic name covering a variety of species including the eagle, vulture, griffon, and kite, according to the wording of the text. Indeed, one scholar clearly distinguishes between a griffon and an eagle in these words: "While the eagles and other birds are content with lower elevations, the griffon alone selects the stupendous gorges of Arabia, Petraea, and the defiles of Palestine."[1] Evidence for this is found in the Bible: "Does the *nesher* mount up at thy command and make her nest on high" (Job 39:27) and parallels. For this reason, the *nesher* is called the king of the skies as it reaches the highest parts of the loftiest rocks.

One medieval moralist conveys this lesson: "The image of the eagle is to teach us this lesson, that even as the eagle soars higher but swoops down, so should scholars act. Though they be brilliant they should not be conceited in the presence of their teachers, but listen humbly so that they may learn from them."[2]

On the other hand we find this intriguing comment of Radak on Isaiah 40:31 who quotes Saadiah Gaon to the effect that the *nesher* rises higher every ten years and drops its feathers, which are burnt through the rays of the sun.

The Hebrew word *nesher* is derived from the root *nashar*, "to drop," but in the Piel can mean "to tear." These meanings reflect the actions of the eagle: his beak is strong and hooked, his claws are long and sharp, he flies very high, and with his keen sight he has a clear view of the landscape beneath and never misses an opportunity to swoop down swiftly, seize, and tear his prey so that it falls apart, "as the vulture that swoops on the prey" (Job 9:26).

Allied to its ability to achieve great heights, the eagle is also famed for its swiftness; "our pursuers are swifter than the eagles of the heaven" (Lamentations 4:19). And the Mishnah advises us symbolically to be "swift as an eagle to do the will of your Father Who is in heaven" (*Pirkei Avot* 5:23).

This important characteristic of the eagle is underscored in the Torah where we find in beautiful poetic imagery the love of God for Israel being compared to the love of an eagle for its young: "You have seen what I did to the Egyptians and bore you on eagles' wings and brought you unto Myself" (Exodus 19:4). In the words of Rashi, who quotes the *Mechilta*: "As an eagle which bears its fledglings upon its wings, Scripture uses this metaphor because all other birds place their young between their feet since they are afraid of other birds that fly around them, but the eagle fears none except man who may shoot arrows at it, as no bird flies above it. For this reason the eagle places its young upon its wings saying: 'better that the arrows and missiles strike me rather than my young.' "

The full import of the above passage is evident from Rashi's comment on Exodus 14:19 where we learn that the angel of God and the pillar of cloud divided the camp of Egypt from that of Israel, causing God to receive the arrows of the Egyptians. In other words, God in His abundant love and kindness to Israel diverted the missiles directed against His beloved people and, like the eagle, accepted them under the wings of the Divine presence.

This care and consideration of the eagle toward its young are further portrayed in Moses' farewell song: "as an eagle that stirs up its nest, hovers over its young, spreads abroad its wings, takes them beneath them on its pinions" (Deuteronomy 32:11). Here Rashi quotes the *Yalkut*, which informs us that God's loving concern and deep attachment to Israel are compared to the eagle that does not suddenly enter the nest but flutters its wings between the branches and bushes, gently awakening its young. In this manner the young are prepared to fly, and they obediently submit to the training and discipline that eventually take them to the lofty heights and they become independent.

The innate affection of the eagle is not restricted only to its own fledglings but is also extended to those of other birds, as we learn from a remarkable statement in the Talmud. The Rabbis distinguish between the eagle and the wild goat; the latter is heartless toward her young. When she crouches for delivery she goes up to the top of a mountain so that the young shall fall down and be killed, but God prepares an eagle to catch it in his wings and set it before her, and if he were one second too soon or too late it would be killed (*Bava Batra* 16a–b).

Yet another confirmation of the care of the *nesher* toward its young is furnished by Job, who declares that "she dwells and abides on the rock upon the crag of the rock and the stronghold" (Job 39:28). What was the purpose of the stronghold? If perchance a strong bird did reach the crag of the rock it would find the stronghold impenetrable. Thus, the eagle would make it virtually impossible for any harm to befall the young who were enclosed in the stronghold.

The Bible makes reference not only to the care of the young but also to the eagle in old age. The Psalmist suggests that the eagle retains its vitality and alertness even in old age: "Who satisfies your old age with good things, so that your youth is renewed like the eagle" (Psalms 103:5). We know from nature study that the eagle can live to more than a hundred years.

The fifth commandment enjoins us to honor our father and our mother, that our days may be long. We normally associate this commandment with human beings, but it is very tempting to suggest that it might refer also to the animal world. The eagle showers love and affection on its young, and in return the young are respectful and obedient to the old, and hence grow up to enjoy longevity.

In an entirely different context, the Book of Proverbs bids us not to covet riches, "for riches certainly make themselves wings, like an eagle that flies toward heaven" (Proverbs 23:5). The metaphor teaches us that as the eagle rises higher and higher until it is almost out of sight, so our riches, the wheel of fortune, may suddenly turn and we have lost all our wealth.

In one instance there is a reference to the bearded vulture: "Enlarge your baldness as the vulture" (Micah 1:16). It appears that the vulture moults its feathers sooner than the other birds, and consequently it has the semblance of baldness.

In addition to the *nesher*, which incorporates the griffon vulture, two other species of this family are specifically named in the Torah: *peretz* and *ozniyah*. These species follow *nesher* in Leviticus 11:13 and Deuteronomy 14:12. The authorized and revised versions of the English Bible translate *peretz* as gier-eagle but the Jewish Publication Society edition calls it bearded vulture. The Hebrew word *peretz* is derived from a root meaning to divide or break into pieces, hence the Latin name *ossifrag*, the bone crusher. This name characterizes the practice of this bird, which at times snatches animals that it then carries high and casts against the rocks, crushing their bones.

The Rabbis assert that both the *peretz* and *ozniyah* are not found in inhabited settlements (*Chullin* 62a).

The *ozniyah* is translated as "vulture or osprey." Some call it a sea eagle because it seizes the fish near the surface of the water with its strong talons. It is a powerful bird and is known in the Mishnah as *oz*, meaning "mighty and strong." The rabbis inform us that implements plated with metal were made from its powerful wings (*Kelim* 17:14). The derivation of *ozniyah* is unknown.

In antiquity the eagle was the chief standard of the Roman legions, and the word *nesher* symbolized Rome. Thus, we learn that scholars arrived from Tiberias who had been captured by an eagle—that is, a Roman (*Sanhedrin* 12a), and in another instance we read that an oath was taken "by the Roman eagle" (*Pesachim* 87b).

Proverbial Sayings

It is a disgrace for an eagle to perish in its gilded cage (Z. Shneor).

As an eagle that stirreth up her nest, hovereth over her young . . . the Lord alone did lead him . . . (Deuteronomy 32:11).

Elephant

Modern commentators state that the elephant is not specifically mentioned in the Bible, but the tusk of the elephant is found several times. Rashi on 1 Kings 10:22 renders *shenhabim* as the tusk of the elephant, and in the parallel passage, 2 Chronicles 9:21, Rashi clearly states that the word "*shenhabim*" (ivory) means elephant. Both Targum Jonathan and Radak translate this word as "tusk of elephant."

The first direct reference to elephants is in the Book of Maccabees where we read that Antonius Epiphanes overran Egypt "with a great multitude, with chariots, elephants, horsemen, a great navy" (1 Maccabees 1:17). From this passage we learn that the elephant was widely used in the wars of Antiochus against the Jewish people. In the first campaign against the Jews, 32 elephants were employed. They were the spearhead for 20,000 horsemen and 100,000 foot soldiers (1 Maccabees 6:30). Each elephant led the way for 1,000 foot soldiers and 500 picked horsemen (1 Maccabees 6:35).

Eliezer, the brother of Judas Maccabeus, was crushed to death by an elephant he had already slain (1 Maccabees 6:46). The elephants were well trained, and a special officer, a master, was set over them (2 Maccabees 14:12). Before the final battle with Judas, the Apocrypha refers to the savageness of the elephants, which suggests that they were given intoxicating drinks to make them more daring (2 Maccabees 15:21).

In Mishnah and Talmud the elephant is called *pil*. Jastrow suggests that it was originally *naphil* (plural *nephilim*), meaning "giant" (see Genesis 6:4 and Numbers 13:33).

The elephant is a big, tough, heavy animal and was generally classified as a giant animal. Indeed it weighs more than three tons. The massive elephant is contrasted with the small mosquito, which, once it enters the trunk of the elephant cannot be shaken off easily (*Shabbat* 77b). (See Rashi.) In the introduction to *Seder Zera'im* Maimonides also refers to two extremes in the expression "from the elephant to the worm."

The Talmud offers a different interpretation for the word *pil*. The Rabbis declare that if one sees an elephant in a dream, wonders (*pela'im*) will be wrought for him; if several elephants are seen, wonder of wonders will ensue (*Berachot* 56b, 57b). From the above statement we learn how keen the Rabbis were to emphasize the wonderful and miraculous nature of God's creations on earth. The wise and beneficent Creator of the universe has endowed every creature with wonderful instincts. The Rabbis, therefore saw *pela'im*, "wonders," in the *pil*.

This sense of wonder and mystery was responsible for a special blessing: "Blessed art Thou . . . who changes or varies the creatures." Our rabbis have taught that "he who beholds an elephant or ape says, 'Blessed . . . who varies the forms of Thy creatures' " (*Berachot* 58b; see also *Shulchan Aruch, Orach Chayyim* 225:8).

In one sense the elephant is different from other animals; it does not have normal feet capable of running, but moves very slowly. We thank God for creating each creature with its peculiar limbs, which are used to advantage when others are lacking. The elephant cannot lower its head so therefore has a long nose or trunk with which it receives its food and drink. He can live to more than 100 years. The elephant is wise and subtle and eats only one meal a day. Its roar is similar to that of a lion. It is very strong and can pick up tiny things with its trunk.

This blessing, as recorded in Singer's prayer book, has the following commentary: "On seeing strangely formed persons such as giants or dwarfs."[1] Such a description is misleading for it would suggest that the blessing is said only over freakish beings, but this is incorrect, for the Talmud incorporates into the blessing the elephant and the monkey, and these animals are not freakish.

We cannot resist referring to a most interesting detail regarding elephants recorded by Richard Eden in 1577. He tells us that the elephant battles with dragons as a matter of habit but is so chaste that having once mated with a female, the male elephant "never toucheth her," a most certain caution against lust.

Proverbial Saying

Rava said, "A man is never shown in a dream a golden date palm or an elephant entering the eye of a needle" because he never thinks of such things (*Berachot* 55b).

Fish

The Hebrew word for fish is often found in the Bible both in the masculine form, *dag*, and in the feminine, *dagah*, and both are derived either from the root *dagah*, "to multiply or increase" (Genesis 48:16) or *doog*, a denominative verb meaning "to catch fish" (Jeremiah 16:16): "Behold I will send for many fishers . . . and they will fish them."

Apart from *tanin*, *leviathan*, and *rahab*, mythical sea monsters at times used symbolically against nations, the Bible employs the word *dag* as a generic term referring to all types of fish. Even the story of Jonah does not specifically mention the name of the fish that swallowed Jonah, but simply informs us that it was a *dag gadol*, "a big fish," either a whale or a shark, both of which could hold a human being.

The first mention of the fish in the sea in the Bible is found in Genesis 1:26. The Patriarch Jacob blesses Ephraim and Menasseh with the words *vetidgu larov*—and let them grow into a multitude (Genesis 48:16). Rashi comments: "Like fish which are fruitful and multiply."

That the waters of the sea abound with shoals of fish is confirmed by the Torah, which records that the Israelites in the wilderness pined and craved for the fish they acquired in Egypt *chinom*, "for nothing" (Numbers 11:5). The medieval commentator, the Abravanel, explains the word *chinom* literally, adding, "The Nile waters overflowed; all one had to do, was to dig a hole which was filled by the waters of the river.

When the Nile receded, the fish remained in the pits. In this way, they ate them free of charge." The Great Sea, the Mediterranean, also teemed with many fish, as we learn from the prophet Ezekiel: "and there shall be a very great multitude of fish" (Ezekiel 47:9–10).

An interesting parallel to the link between fish and the propagation of species is found in the term *nun* (Aramaic, *nunah*). In Psalm 72:17, the words *yenon shmo* are translated by the B. D. B. Hebrew Lexicon as "let his name be propagated." The same word, *nun*, is used in the Talmud to mean "fish" as in *Nedarim* 54b: "The succession of letters *Nun*, *Samech*, *Ayin* serves as an intimation 'fish is a remedy for the eyes.' "

The Talmud confirms that Babylonia with its many rivers and ponds was well stocked with fish. Thus the rabbis record that on a certain day, "everybody was engaged in fishing and they brought in the fish on Pesach week, and the sage Rava allowed them to put the fish in salt (because there was more than enough for the people's requirements)" (*Mo'ed Katan* 11a).

In another instance, we learn that if one dams a pond from a stream on the eve of a festival so that no fish can come in and on the festival morning he finds fish therein, they are permitted to be eaten during the festival as the fish must have been in the channel before the festival began (*Betzah* 25a).

The popularity of fish was so marked that it even entered the vocabulary of Jewish law and custom. This may be due to a favorite expression of the rabbis, who often refer to the *yam haTalmud*, the sea of the Talmud. As the oceans are unfathomable, so are the words of Torah. One cannot fully plumb the depths of rabbinic wisdom. If this be true of the waters of the ocean, the fish too must receive recognition, and serve symbolically as a message for mankind. For example, the *Tashlich* ceremony is observed on the afternoon of the first day of Rosh Hashanah (if it is a weekday), and young and old congregate at the edge of a river or stretch of running water that preferably contains fish. This teaches us that as fish are caught in nets, so human beings succumb to the pitfalls of life. This custom reminds us of the classical parable of Rabbi Akiva, who, undaunted by the harsh decree of Hadrian not to teach Torah in public, continued to spread the exposition of Torah. When his colleagues remonstrated with him not to endanger his life, he replied with the following parable:

"A fox saw fish moving to and fro in the water. The sly fox suggested to the fish that they leave the insecurity of the water and come on dry land where they would be safe from the nets of fishermen. The fish promptly replied: 'If we are in danger in the water which is our natural habitat, how much more so on dry land where we would automatically perish.' The Torah is our life element," continued Rabbi Akiva, "we cannot survive without it" (*Berachot* 61b).

Rabban Gamliel the Elder analyzed the characteristic qualities of his disciples by comparing them to different types of fish. "An unclean fish is a poor youth who studies Torah and is without understanding. A clean fish is a rich youth who studies Torah and has understanding. A fish from the Jordan is a scholar who studies Torah and has not the talent for give and take. A fish from the Great Sea (Mediterranean) is a scholar who studies Torah and has the talent for give and take" (*Avot d'R. Nathan* 40:9). In this connection it should be noted that one schoolteacher who had a fish pond attracted his pupils to attend lessons regularly by rewarding them with fish dishes (*Taanit* 24a).

As fish were very familiar to every segment of the people, including scholars, the question arises, Does the Bible enlighten us on how the fish were caught? In fact, the Bible mentions a variety of implements used for fishing: the *choach*, "hook or ring" in the jaw of a large fish such as the crocodile (Job 40:26); *chakah*, "fish hook" (Job 40:25); *michmeret*, "fishing net" (Isaiah 19:8, Habbakuk 1:15); *cherem*, "net" (Micah 7:2, Ezekiel 26:5, 32:3); and *tziltzal dagim*, "fish spears or barbed irons" (Job 40:31).

The Talmud too enumerates a variety of fishing tackle by name. We also learn that it was permitted during the Festival week to fish with an angle in the Sea of Tiberias, and the rabbis add that it was allowed to fish by means of nets and traps (*Bava Kamma* 81b). All this points to the fact that many Jews practiced fishing. But the Talmud mentions one fisherman by name—Alda—who is reported to have told his teacher

Rav, "Broil the fish with its brother (salt), plunge it into its father (water), eat it with its son (sauce) and drink after it, its father (water)" (*Mo'ed Katan* 11a). In a striking passage an early example of fishing is provided by *The Testament of the Twelve Patriarchs*, which records the last words of the twelve sons of Jacob, written probably at the end of the second century B.C.E. Zebulin proclaims: "I was the first to make a boat to sail upon the sea . . . and I let down a rudder and I stretched a sail and sailed along the shore catching fish for the house of my father and through compassion I shared my catch with every stranger—if there were aged I boiled the fish and dressed them well and offered them to all men and the Lord satisfied me with an abundance of fish."

Because fish were in plentiful supply and meat was considered a luxury, beyond the reach of the poor, fish became the staple diet of Jews. To this day, we connect fish with celebration of the Sabbath and Festivals as we see from the following passage: he who possesses a *manah* (a hundred shekels) should buy a measure of vegetables for his pot, if he possesses ten *manah*, he should buy a quantity of fish for his pot, if he possesses fifty *manah*, he may buy a quantity of meat for his pot (*Chullin* 84a). The following story illustrates the importance of eating fish on the Festivals:

A pious man in Rome was accustomed to honor the Sabbath and Festivals. On the afternoon before the Day of Atonement, he went to the market and found only one fish for sale. Now the governor's servant was standing there, and they bid against each other for the fish. Eventually, the Jew bought it at a denarius a pound. At dinner, the governor asked why there was no fish, and when he was informed about the Jew and his purchase, the governor accused the Jew of having a hidden treasure belonging to the king. The Jew pleaded that as the Day of Atonement is the Sabbath of Sabbaths and must be honored, he felt impelled to spend much money to acquire the fish. Thereupon, the governor acquitted him. There is a happy ending to the story. The Almighty rewarded the Jewish tailor and prepared for him a precious pearl in the fish, and on the proceeds of the pearl the Jew lived happily for the rest of his life (*Pesikta Rabbati* 11a).

Regarding the honoring of the Sabbath with food, the rabbis are of the opinion that a poor man may not accept charity even in order to make the Sabbath a delight. In answer to the query, what food should he purchase in honor of the Sabbath, the rabbis reply, "fish-hash" (*Pesachim* 112a).

Though the word *dag* is generally used for many types of fish, the rabbis do specifically mention by name several types of fish. There is the *shibbuta*, probably mullet or carp; Rava salted a *shibbuta* for the Sabbath (*Kiddushin* 41a). *Kilbit* is supposed to be a stickleback (*Avodah Zarah* 39b). The rabbis also mention a group of small fish, some of which cannot be identified, such as *sultanith* (a kind of anchovy), *aphiz*, *colias*, *scomber*, swordfish, *anthias*, *tunny*, *zahanta*, *shefarnuna*, *kedashnuna*, and *kevarnuna* (*Avodah Zarah* 39a).

We cannot conclude this review without some reference to the Jewish dietary laws according to which only fish that have fins and scales are permitted to be eaten (Leviticus 11:9, Deuteronomy 14:9). The Talmud deduces that when a fish has scales, it invariably has fins and is therefore permitted. Since this principle was established, new rivers and lakes have been discovered with countless species of new fish. Nonetheless not a single fish has been found having scales without fins. It is also reported that the only fish that fishermen regard as nonpoisonous are those with fins and scales.

It is not our purpose to dilate on the dietary laws; we shall refer only to one historical fact which proves the efficacy of traditional law. "When David Rahabi came to India and found some people whom he rightly thought to be Jews though they were scarcely distinguishable from their Indian environment, it was not the *Shema* which proved that they belonged to the House of Israel but the fact that they eschewed fish that lacked fins and scales."[1] The same authors quote William Radcliffe, who in his book *Fishing from Earliest Times*, blames Jews for lacking the sporting spirit; they caught fish by the net, and they did not play with the rod. Loewe adds these significant words: "the word 'hook' occurs in the Bible only as a metaphor of cruelty or as an instrument used by foreigners. In Rabbinical times the hook which entered the mouth of the fish typified cruelty and with it was compared the terrible disease of croup, which similarly attacks and chokes infants. This was the 'evil net' of Ecclesiastes 9:12."[2]

Moralistic Stories

Rabbi Bar Chanan, "Aesop" of Talmud, describes in one of his fabulous tales the sad plight of a group of sailors who were eager to escape the uncertainties of a storm-tossed boat for a nearby island. As soon as they set foot on land the island began to sink. They discovered too late that they had abandoned the relative safety of a boat for the whims of the back of a whale. Many have abandoned the safe but storm-tossed ship of the Torah for shifting sands of specified modern knowledge.

Socrates, the Greek philosopher, was asked why he insisted that every aspiring disciple must first look at the community pond and tell him what he sees before he is admitted into the fellowship of the initiated. The student answered that those who looked into its blue waters and said they saw themselves were rejected. Only those who said they saw fish swimming were accepted.

Proverbial Saying

Like fish like men–the greater swallow the smaller (*Avodah Zarah* 4a).

Fly

According to the B. D. B. Hebrew Lexicon, the Hebrew word for fly, *zevuv*, is "something which moves to and fro in the air." The word is found in Ecclesiastes 10:1: "Dead flies make the ointment of the perfumer fetid and putrid, so a little folly outweighs massive wisdom." The Targum is revealing, for it equates the fly with the evil inclination, sin: "The evil inclination which dwells at the gates of the heart (Genesis 4:7) is like a fly and causes death in the world, because the wise man befouls himself at the time that he sins."

Another reference to *zevuv* in the Bible is found in connection with the idol Baal. "He (Amaziah) sent messengers and said unto them: Go, enquire of Baal-Zevuv the God of Ekron . . ." (2 Kings 1:2). Baal-Zevuv, literally "Lord of the Fly," was a Philistine deity believed to control the movement of flies to and from a locality. In a hot country like *Eretz Yisrael*, flies may be the cause of serious epidemics (Soncino Bible).

In antiquity the fly was venerated as a deity by many people. In Ekron, one of the five Philistine cities, a magnificent temple was erected in honor of Baal-Zevuv, the Lord of the Fly.

The prophet Isaiah employs the fly to symbolize the armies of Egypt, which are compared to a swarm of flies: "And it shall come to pass on that day that the Lord shall hiss for the fly that is in the uttermost part of the rivers of Egypt" (Isaiah 7:18).

In Exodus 8:17 the plague of *arov* is sent upon Egypt; this word is generally translated as a swarm of flies. The Septuagint renders it as a stinging fly, whereas Philo calls it a dog-fly, a ferocious insect (*Vita Mosis* 1:23); compare Psalms 78:45 and 105:31 where the same word, *arov*, is used.

An indirect reference to the fly is found in the story of Joseph where we learn that the butler of Pharaoh sinned against the king. According to the Midrash, quoted by Rashi, the butler sinned by allowing a fly to be found in the king's cup of wine. For this lapse of duty he was incarcerated and met Joseph who interpreted his dream. How often is the fate or destiny of an individual or nation determined by an insignificant object; in this case, a fly! It was the fly that triggered off a concatenation of events that eventually led to the meteoric rise of Joseph to become viceroy of Egypt.

The Talmud and Midrash record some interesting data regarding the fly. Everything that God created has a purpose. Thus, applying a crushed fly to one who is stung by a hornet acts as an antidote (*Shabbat* 77b). "Why is the tail of the ox long?–When it grazes in the meadow, it can beat off the flies with its long tail" (*Shabbat* 77b).

The Egyptian fly that is mentioned in Isaiah 7:18 is so dangerous that it may be killed on the Sabbath (*Shabbat* 121b). The Mishnah refers to the spreading of a cloak over animals to protect them from flies (*Parah* 2:3). Curtains were hung over beds as a protection against flies (*Sukkah* 26a). Flies were considered such a nuisance that prayers were ordained for their removal (*Taanit* 14a).

In *Pirkei Avot* 5:5 the rabbis assert that a fly was never seen in the slaughterhouse of the Temple in Jerusalem. If we take into consideration the exceptional heat in *Eretz Yisrael* together with the fact that there was no refrigeration such as we enjoy today, it was a miracle that no fly was seen in the abbatoir. Maimonides was of the opinion that the flies were driven away by the smoke of the incense, whereas Rashi maintains that the absence of flies was due to the tables of marble; in *Mishnah Shekalim* we learn that there were eight tables of marble in the Temple (*Shekalim* 6:4). In another passage, we learn that the sons of Eli were negligent in that they left the juicy parts of the sacrifice exposed to the flies (*Yalkut Shemuel* 8b).

The story of the wicked Titus is well-known. He blasphemed God, and a fly entered his nostrils and it knocked against his brain for several years till his death (*Gittin* 56b).

Among the accusations devised by Haman against the Jewish people before King Ahasuerus was the following: "They eat, drink and despise the throne. If a fly falls into a cup of one of them, he throws it out and drinks the wine, but if my Lord the King were to touch his cup, he would dash it to the ground and not drink of it" (*Megillah* 13b).

The above quotation reminds us of the midrashic passage in which the rabbis portray the drinking habits and temperaments of different people. There is one type who, if a fly falls into his cup, flicks out the fly and drinks the contents. This is the case with an ordinary man who, if he sees his wife gossiping with her neighbor, leaves her alone. There is another type who, should a fly flutter over the surface of his cup, will seize the cup and pour out its contents without tasting it; this is an evil quality in people. There is a type of person who, if a fly settles on the surface of his cup, takes it and puts it down untouched. There is a type of person who, if a dead fly falls into his cup, takes out the dead fly, sucks it, and then drinks; such a person is a wicked person (*Numbers Rabbah* 9:8).

Proverbial Sayings

If man does worthily, they say to him, "You were created before the angels"; if not, they say to him, "The fly and the worm were created before you" (*Genesis Rabbah* 8:1).

Never saw a fly pass by his table (*Berachot* 10b).

Fowl and Fowling

Fowl is a general term for any winged creature used for Jewish consumption, such as a cock, hen, and bird. The fowler is one who hunts either for sport or food. As hunting for sport is frowned upon and discouraged in Jewish life, this essay deals with fowling or trapping only for food.

However, nonkosher animals, for instance those animals that are covered with fur, were trapped as well. Thus we learn that cats were trapped because of their fur, which had commercial value (*Bava Kamma* 80b). We are also informed that the skins of birds and the hides of animals were used by Jews for writing, and bird skins were used for the making of scrolls and *tefillin*. Skins of fish were also used for parchment. Indeed the Talmud discusses whether the skins of fish could be used for scrolls or *tefillin* (*Shabbat* 108a).

Fowling seemed to be practiced by a goodly number of Jews in Babylonia, and the variety of instruments employed by fowlers is evident in the biblical period, which proves that fowling and trapping were not a rarity. Thus we see the *Reshet*, a net to trap birds, in Proverbs 1:17. Hosea also mentions a net: "Even as they go, I will spread my net upon them: I will bring them down as the fowls of the heaven" (Hosea 7:12).

In Amos 3:5 we have *pach* and *mokeish*: "Will a bird fall in a snare upon the earth where there is no lure for it? Will a snare spring up from the ground and have taken nothing at all?" (Amos 3:5). A snare is a device with a spring attached that clamps down upon the unsuspecting victim attracted by the bait. In Psalm 140:6 *pach* is followed by *chevel*, a "cord or noose" that is concealed in the ground to catch the leg (compare Job 18:9 and 10).

In the Talmud fowling was known as *rishba* and was used as a family name; thus, Joseph Rishba–Joseph the fowler (*Shabbat* 130a; *Chullin* 116a), Tabi the hunter (*Shabbat* 17b), and Tabuth the trapper (*Sanhedrin* 97a, *Taanit* 10a). In the fourth century there is mention of a trapper called Papa ben Abba (*Chullin* 54a).

The most fantastic hunter of all must surely be Rabbi Yona ben Talifa, who slaughtered with his arrow a bird in flight, and it was in accordance with the laws of *shechitah*. He must have been an excellent marksman to have accomplished such an amazing feat. The talmudic sage, Rava, testified that the bird was correctly slaughtered according to Jewish ritual.

In addition to the instruments for catching birds that are found in the *Tanach*, a few are mentioned in the Talmud. The *izla* is a web in which the knots are close or far apart (*Chullin* 51b). There also is the *Devok*, "a stick," the fowler's rod with the board on top smeared over with glue in order to catch birds (*Chullin* 52a). In *Shabbat* 80a there is mention of a line board of a hunter's rod. Finally, a *shutah*, a trap made of a framework or a block, is mentioned in *Bava Kamma* 117a in connection with a dispute between two men about a trap and the animals caught in it.

It is therefore apparent that both in the biblical and the talmudic period Jews were not only bird watchers, but bird catchers.

Proverbial Sayings

A bird of the air shall carry the voice (Ecclesiastes 10:20).

Like a bird sticks to its kind, so truth to its seekers (*Ben Sira* 27:9).

There are people who believe they know the fowl because they saw the egg from which it emerged (Heinrich Heine).

Fox

It is conjectured that the Hebrew word for fox, *shual*, is connected with *shaal*, meaning "hollow": "who has measured the waters in the hollow of his head" (Isaiah 40:12; see also 1 Kings 20:10). The same word is used of a narrow road or path: "Then the angel of the Lord stood in the hollow way between the vineyards" (Numbers 22:24). The connection between *shual* and something hollow is that the fox likes holes or narrow paths.

Another obvious association in the above verse is the fondness of foxes for grapes. This is confirmed by the Song of Songs: "Take us the foxes, the little foxes that spoil the vineyards, for our vineyards are in blossom" (Song of Songs 2:15). In the United States fox-grape is the name for a special species of wild grapes.

The first reference to foxes in Scriptures is found in Judges 15:4 where we learn that Samson took 300 foxes, tied them by their tails, placed a firebrand between each, and let them loose in the cornfields of the Philistines. Many commentators suggest that the Bible refers here to jackals and not foxes, because jackals gather in packs and were more numerous than foxes, which usually move around singly.

It should be noted that the jackal and fox resemble each other and are therefore mistaken for one another. And the Soncino Bible informs us that the Hebrew word *shual* is derived from the Persian *shagal*, the origin of the English jackal. When the Psalmist claimed, "They shall be hurled to the power of the sword; they shall be a portion for foxes" (Psalm 63:11), he was probably referring to the jackal, which is known to feed on dead bodies.

An echo of this confusion between fox and jackal is reflected in the Talmud, which records that a fox tore a lamb. When the case came before the rabbis they decided that the lamb was not bitten by a fox, but by a dog (*Chullin* 53a). Some scholars believe that the jackal is an ancestor of the dog.

From Lamentations 5:18, we know that foxes prowl in ruins and desolate places, "for the mountains of Zion which is desolate, the foxes walk upon it." And the Prophet Ezekiel observes that "thy prophets have been like foxes in ruin" (Ezekiel 13:4).

The fox plays a prominent part in rabbinic literature, especially in the *Aggadah*, the section of the Talmud that deals with folklore and legend. Thus it is reported that Rabbi Meir had a collection of three hundred fox fables, but only three are in existence (*Sanhedrin* 38b, 39a). Rashi in *Sanhedrin* 39a combines and summarizes them into one story. However, in addition to these three tales, many more fox fables are scattered throughout the Talmud.

The beautiful Midrash in *Ecclesiastes Rabbah* 5:21 on the verse "as he came forth out of his mother's womb naked, so shall he return" is presented below:

A Babylonian teacher named Geniva compared man in this world to a fox that has found a vineyard surrounded on all sides with a high fence except for a small opening at one point. The fox attempted to enter, but finding the hole too narrow to squeeze through, he began to starve himself for three days until he was thin enough to enter. Once in the vineyard, he indulged himself to such an extent that he regained his weight. In desperation, the fox was compelled to fast another three days to enable him to leave the vineyard. Then he exclaimed: "Oh vineyard, how pleasant are you and how desirable are your fruit, but of what benefit are you to me since I depart from you as thin as when I entered"; such, says Geniva, is the fate of man in this world, as man came forth so he returns.

Rabbi Johanan ben Zakkai too was complimented by the rabbis for including in his studies the fables of the foxes (*Bava Batra* 134a).

The following incident illustrates the absolute faith and optimism of Rabbi Akiva and is worthy of repetition. As Rabban Gamliel, Rabbi Eliezar ben Azariah, Rabbi Joshua, and Rabbi Akiva went to Jerusalem together and reached Mount Scopus, they saw a fox emerging from the Holy of Holies. Rabbi Akiva seemed to be merry, but the other rabbis wept. They asked Akiva why he was merry. Akiva asked them why they wept. They said to him: "a place of which it was once said, 'and the common man that draws near will be put to death' (Numbers 1:51) now becomes the haunt of foxes, and should we not weep?" Said Akiva to them, "Therefore I am merry, for it is written, 'and I will take to me faithful witnesses to record, Uriah the priest and Zechariah the son of Jeberchiahu' (Isaiah 8:2)."

"What connection has Uriah with Zechariah? Uriah lived during the time of the First Temple, while Zechariah lived and prophesied during the Second Temple. But holy writ linked the later prophecy of Zechariah with the earlier prophecy of Uriah. In the earlier prophecy it is written, 'therefore shall Zion for your sake be plowed as a field.' In Zechariah it is written, 'thus says the Lord of Hosts, there shall yet old men and old women sit in the broad places of Jerusalem' (Zechariah 8:4). So long as Uriah's threatening prophecy has not had its fulfillment, I had misgivings lest Zechariah's prophecy might not be fulfilled. Now that Uriah's prophecy has been fulfilled it is quite certain that Zechariah's prophecy also is to find its literal fulfillment." They said to him, "Akiva, you have comforted us" (*Makkot* 24b).

The famous proverb "Be a tail to lions and not a head to foxes" is found twice in the Talmud, once in *Pirkei Avot* 4:20 and once in *Sanhedrin* 37a. But each reference has a different connotation.

The proverb in the Mishnah seems to be associated with leadership. The lion is the king of beasts, the leader and head of the family. The fox is noted for its sly and cunning nature; it is also considered to be a foolish animal. One should remember that it is the nature of the lion to lift up its tail to its head, whereas the fox lowers its head to the ground. Similarly, man should not aspire through cunning and dishonest means to become a leader and in the process bring ruin and destruction on those who trust him. One should rather be a tail to a lion, a loyal and trustworthy follower, but do not become a corrupt and cunning leader.

To understand the full import of the proverb in *Sanhedrin* 37a, we must realize that in the ancient academy of learning, scholars sat in rows and when the head of the row was promoted all scholars moved forward. Therefore, the proverb "better a tail to lions than a head to foxes" conveys the meaning that it is better to be placed at the tail of the first row than at the head of the second row.

In post-talmudic literature, Berachia Hanakdan (19th century) wrote a work entitled *Mishlei Shula'im*, which was a collection of fox fables.

Shibolet shual, one of the substances that can be used to make *matzot* for Passover, are ears of corn that are foxtail in shape (*Pesachim* 35a).

We find that the Egyptians are compared figuratively to foxes. Rabbi Eliezer ben Simon said: "The Egyptians were cunning and therefore compared to foxes. As the fox always looks behind him, so the Egyptians always looked behind them, examining their past" (*Exodus Rabbah* 22:1).

The great chasidic master, the Besht (1700–1760), once said: "The lion became enraged at his subjects, the animals of the forest. He asked the fox to placate the king of the beasts by relating to him an appropriate fable. The fox replied that fear caused him to forget his fables. Hence the beasts were compelled to wait on the lion themselves. In the same manner," said the Besht, "when we approach the High Holidays, the congregation should not depend on their rabbi to pray on their behalf; each one should pray for himself."

Proverbial Sayings

A fox does not die from the dust of its den (*Ketubot* 71b).

They carried away from it only what a fox carries away from a plowed field (*Niddah* 65b).

Be a tail to lions and not a head to foxes (*Pirkei Avot* 4:20).

Frog

The derivation of the word *tzifardaya*, "frog," is uncertain among grammarians, but the rabbis suggest that it is compounded of two words, *daya* and *tzippur*, "a bird of knowledge" (*Yalkut Shimoni* 182). We would add that the frog is a bird with the knowledge of God. This is reflected in several statements in rabbinic literature. In the *Pesikta Vayehi* 66b we read, "He sent them carriers or heralds in the form of frogs." The frog was also favored and referred to as "a messenger of God." It is reported in the name of the sage Samuel that he witnessed an incident in which a scorpion that cannot swim was borne by a frog across the river; the scorpion then stung a man so that he died. Thereupon Samuel quoted the verse "They stand this day according to thy ordinances; for all things are thy servants" (Psalm 119:91). This verse implies that we are all servants ready to carry out the judgments of God; in this instance it was the frog that was the servant of God.

This "favored" frog also appears in a fascinating story recorded in the Midrash at the end of the Book of Psalms. When King David concluded the writings of the psalms he was naturally elated and offered up a prayer of thanksgiving to the Almighty, in the course of which he exclaimed: "Is there a creature in the universe that sings praises and songs to the Almighty as I?" Suddenly a frog presented itself and embarrassed David by asserting proudly: "Do not boast of your achievement. I sing more praises and songs to God than you." The frog was even emboldened to compare himself with King Solomon of whom it is said, "And he spoke three thousand proverbs and his songs were a thousand and five" (1 Kings 5:12; *Yalkut Shimoni* 889).

What warranted the frogs to assert that they sang more praises to the Creator than David? To reply to this question we proceed to the plague that the Almighty brought upon the Egyptians. In Exodus 7:26 we read that God warns Pharaoh that the river shall swarm with frogs that shall go up and enter the homes and even the ovens. In a striking passage the Rabbis portray the frogs as courageous defenders of the faith ready to make the supreme sacrifice to their Maker. They not only sang praises through the croakings that emanated from their mouths but they were also prepared to leave the cold waters of the river that were their natural habitat and plunge into the hot ovens of the people.

They were ready to be burned alive for the sanctification of the name of God. The daring exploits of the frogs were a classic example of divine allegiance, and it was held up as a model for people to emulate. Thus we learn that Hananiah, Mishael, and Azariah, the companions of Daniel, delivered themselves up to the fiery furnace and argued in this manner: "If frogs which are not commanded concerning the sanctification of the divine name, yet they were ready to go into the houses, ovens and kneading troughs of the people, we who are commanded to sanctify the divine name, how much more so" (*Pesachim* 53b).

It may sound strange but according to rabbinic exposition, the plague of frogs seems to have solved a problem that agitated the Egyptians and Ethiopians. Commenting on the verse "Behold I will smite all the borders with frogs" (Exodus 7:27), the rabbis observed that the plague of frogs brought upon the Egyptians was the means of establishing peace. There was a dispute between the Ethiopians and the Egyptians in which each accused the other of encroaching on its borders. When the plague of frogs came, the dispute was settled, as the frogs reached only the borders of Egypt, thus creating a clear line of demarcation between the two nations (*Exodus Rabbah* 10:2).

In another passage in the Bible, it is written, "the frog came up and conquered the Land of Egypt" (Exodus 8:2). The rabbis questioned the singular use of the word *frog*. This gave rise to a difference of opinion between Rabbi Eliezer and Rabbi Akiva; the former said it was one frog that bred prolifically and the latter said one frog croaked for the others to come (*Sanhedrin* 67b) (*Exodus Rabbah* 10:5).

The following note is found in *Midrash Tehillim* on Psalms 78:45. Rabbi Jonathan taught: Wherever an Egyptian sat down there a frog appeared, but what of the houses of marble and stone, how did the frogs get into them? They would break into them by saying they were the emissaries of the Almighty. Thereupon the marble and the stones would split open before them.

It is interesting to record that even in the Halachah, the frog in one sense is "favored." In Leviticus 11:29 there is mention of creeping things whose dead bodies defile by touch, but the frog, which is prohibited as food, does not defile when it is dead (*Niddah* 18a).

Proverbial Saying

And then he sent them criers, that is the frogs (*Pesikta Rabbah* 17).

Gazelle

The gazelle has slender legs that help it run swiftly: "Asahel was as light of foot as one of the gazelles that are in the field" (2 Samuel 2:18). Its large eyes and its elegant body are symbolic of human beauty. As we read in the Song of Songs 2:9, 4:5, and 7:4, "My beloved is like a gazelle, . . . like two fawns that are twins of a gazelle."

The gazelle is a clean animal and known for its tasty meat (Deuteronomy 12:22, 14:5). Because of its popularity it was served on the tables of King Solomon (1 Kings 5:3).

The gazelle is famed for its grace and beauty, and its name, *tzvi*, has entered the vocabulary in many usages. Thus we have "beauty of ornament" (Ezekiel 7:20); *Hatzvi Yisrael*, "Thy beauty O Israel"; *Tzvi tifarto*, "His glorious beauty"; *Tzvi l'tzadik*, "glory to the righteous" (Isaiah 24:16). Tzvi is a popular Hebrew name, and the feminine form, Tzivia, is found in 2 Kings 12:2.

In *Biblical Symbolism*, Farbridge writes, "The beauty of the gazelle is a favorite term of comparison with Oriental poets."[1] In Proverbs 5:19 a wife is compared to "the loving hind and pleasant roe." The modern Arab expresses the beauty of the woman he loves by comparing her to the black-eyed gazelle. We can thus understand why on Phoenician gems the gazelle is figured along with the star as a symbol of Astarte.[2] The gazelles of Mecca were probably connected with the cult of Aluzza, who is commonly identified with Aphrodite. Furthermore, the gazelles are sacred symbols in South Arabia in connection with Ashtar worship, and W. Robertson Smith relates an instance in which a South Arabian tribe found a dead gazelle, washed and buried it, and mourned for seven days.[3]

The Talmud describes the *tzvi's* horns, noting that the flesh protrudes into the covering outer sheath (*Bikkurim* 10b). The Talmud also hints at the probable crossbreeding of the gazelle with the goat (*Chullin* 132a). The hybrid offspring, labeled by some authorities "*koy*," is the offspring of a goat and a gazelle (*Chullin* 80a).[4]

In *Pirkei Avot* 5:24 the Mishnah cautions us to run as swift as a gazelle. Rabbi Jonah interprets this passage in this manner: "Let men hasten like the gazelle and not grow weary. In general when people run they get tired, but if they run in order to carry out a commandment, they will not tire." In the Midrashic comment on this verse, just as the gazelle leaps from place to place and from tree to tree, so God jumps and leaps from synagogue to synagogue to bless the children of Israel. (*Numbers Rabbah* 11:3).

A further interesting reference is found in a comment on Song of Songs 8:14: "Be thou like a gazelle." The Rabbis add, "As a gazelle when it sleeps has one eye open and one eye closed, so when Israel fulfills the will of God He looks on them with two eyes, but when they do not fulfill the will of God, He looks on them with one eye" (*Midrash Shir Zuta*).

The land of Israel is compared to a gazelle (deer): as the skin of the gazelle (when once taken off) cannot cover its body so cannot . . . Israel contain its fruits (*Gittin* 57a).

Proverbial Saying

He has put his money on a gazelle's horn (*Gittin* 58b).

Goat

The goat is often mentioned in the Bible and is classified under a variety of names. The most common name is *aiz* (plural, *izim*), as found in Leviticus 3:12 and other verses. The word is probably derived from *oz*, meaning "strength." Confirmation for this derivation is found in the Talmud where we learn that the goat is distinguished for its strength among the small cattle (*Betzah* 25b).

Aiz is also found in combination with other words, as in *seh izim* (Deuteronomy 14:4, Exodus 12:5). This association of sheep and goat possibly goes back to an early period when the goat was as popular as the sheep, as both were found in many homes. At a later stage sheep became more numerous than the goat.

Another combination is *gedi izim*–the kid of the she-goat (Genesis 27:9). The *gedi* plays an important role in the Jewish dietary laws, for the verse "thou shalt not boil a kid in the milk of its mother" (Exodus 23:19) is the basis for the prohibition of mixing meat and milk.

Se'ir izim is the buck of goats (Genesis 37:31). *Se'ir* is connected with *se'ar*, meaning "hair," and reminds us of the English word *goatee*, which is defined as a beard trimmed in the form of a tuft hanging from the chin, resembling that of a he-goat.

Maimonides in his *Guide for the Perplexed* sheds light on the word *se'ir*. He observes that he-goats were always brought as sin-offerings by individuals and by whole congregations on festivals, new moons, and the Day of

Atonement, because most of the transgressions and sins of the Israelites were sacrifices clearly to spirits (*se'irim*, literally goats) as is stated in Leviticus 17:7. Our sages, however, explained the fact that goats were always the sin-offerings of the congregation as an allusion to the sin of the whole congregation of Israel, for in the account of the selling of the pious Joseph we read, "And they killed a kid of the goats" (Genesis 37:31).[1]

Another name used in the Bible is *atoodim*, young he-goats (Genesis 31:10). The word is derived from the root *atad*, to be ready, for the goat is known to be very skillful in evading its pursuer, whether it be man or beast.

Yael, the mountain goat, is also called ibex. Incidentally a kosher crossbreed of the goat and the ibex–goatex–is now making its debut in restaurants and meat shops. Goatex is bred by Kibbutz Lahav in the Negev. The new breed has been named *Yaez* from the Hebrew *Yael* (ibex) and *Ez* (goat). Although Judaism has never approved of intermarriage, even in the animal kingdom–"Thou shalt not let thy cattle gender with a diverse kind" (Leviticus 19:19)–Eliyahu Katz, chief rabbi of Beersheva, has given his official approval to this new breed. It should be noted that since there is a close similarity between the goat and the ibex, it is not a mixed breed.

It is interesting that the strong horn of the *yael* was made into a shofar and also used to announce the Jubilee year (*Rosh Hashanah* 3:3). The word *yael* is connected with *alah* (ascend, climb). It is found in Psalm 104:18–"the high mountains are for the wild goats"–and Job 39:1: "knowest thou when the wild goats of the rock bring forth?" The wild goats are good climbers that reach the highest cliffs of the rocks.*Tzafir*, "he-goat" (compare Daniel 8:5) is an Aramaic word that corresponds to the Hebrew "*se'ir.*" It is a symbol for Alexander the Great, chosen because the he-goat figures in legends of the house of Macedon (Soncino Bible).

Tayish is another word for he-goat (Genesis 30:35); the root of the word is uncertain. In the Book of Proverbs 30:31 the he-goat is reckoned among those that are stately in their march.

Apart from the biblical names mentioned above, the Talmud refers in an obscure passage to the Circassian goat, which in the words of the rabbis is a goat with hooks with which one threshes.

Jastrow explains that the front teeth of the sledge are shaped like goat's horns (*Avodah Zarah* 24b).

The goat was a key element of the sacrificial ritual that took place in the Temple, and today is included in the Yom Kippur liturgy. It is derived from Leviticus 16:7: "And he shall take the two goats and set them. . . ." This ritual has given rise to the word *scapegoat* (the goat that escapes into the wilderness). Incidentally, the word *scapegoat* is not mentioned in the Torah, but was coined by Christian theologians who incorporated it in their translation of the Bible. This word, thus, has a Christological connotation and emphasizes vicarious atonement, which is totally alien to Jewish thought.

We follow the interpretation of Rabbi S. R. Hirsch, on Leviticus 16:10. He explains that according to the Talmud the goats were identical in every respect: one went directly to the sanctuary, and the other (*azazel*) was dispatched to the wilderness. This ritual is interpreted to convey a vital message about the freedom of will with which God has endowed us; we can either soar to the heights of spiritual ecstasy (the sanctuary), or we can descend to the depths of depravity and degradation (the wilderness). We are the masters of our destiny, and it is our privilege and challenge to choose the right path. The hallmark of Judaism is not vicarious atonement, but individual responsibility.

The characteristics of the goat made it both a liability and an asset: the domestic goat was very mischievous and destructive. The Talmud records a number of instances in which the goat causes much damage to crops. Thus in one case, the damage increased to such a degree that it led to litigation, and the court ruled that if after warnings the owners failed to prevent their goats from roaming the streets and fields, damaging the produce, the goats would be seized and slaughtered and the meat offered to the butchers to sell (*Bava Kamma* 23b). In another passage, the goat is referred to as a robber because it tends to pasture in other fields and robs their rightful owner of his crops (*Bava Kamma* 80a).

To counterbalance the disadvantages of goats, we should remember that the goat is a clean and kosher animal, and its meat is wholesome.

The rabbis report that a person who suffered from heart disease was recommended by a doctor to drink warm goat's milk every morning.

In another passage in the Talmud we read that Rami ben Ezekiel once paid a visit to Bnai Brak where he saw goats grazing under fig trees while honey was flowing from the figs, milk ran from them, and these mingled with each other. "This is indeed," he remarked, "a land flowing with milk and honey" (*Ketubot* 111b).

The goat was noted not only for its nutritional properties, for we learn from the Mishnah that the *shofar* blown in the Temple at the New Year was made from the horns of a wild goat (*Rosh Hashanah* 3:3). However, the greatest asset of the goat in biblical days was perhaps its hair, which was used for the making of the curtains of the Tabernacle: "And thou shall make curtains of goats' hair for a tent over the Tabernacle" (Exodus 26:7). Commenting on the verse "and all the women whose hearts stirred them, spun the goats' hair" (Exodus 35:26), Rashi, quoting the Talmud (*Shabbat* 74b), says, "This required extraordinary skill for they spun the hair off the backs of the goats while they were alive."

The same theme is discussed in this Midrash: God said to Israel, "Make Me a dwelling (Exodus 25:8) for I desire to dwell amid My sons." When the ministering angels heard this, they said to God, "Why wilt Thou abandon the creatures above and descend to those below? It is Thy glory that Thou should be in heaven." But God said, "See how greatly I love the creatures below that I should descend and dwell beneath the goats' hair." Hence it says: "Make curtains of goats' hair for the Tabernacle" (Exodus 26:7) (*Tanchuma Terumah* 9).

In the dark Middle Ages the Jew was often portrayed as a demon, evil spirit, and the devil himself. But the devil has no body or form, and so the early Christians chose the innocent domesticated goat to act as the devil (perhaps because of the goat's beard.)

Whatever the reason, the Jew was identified as the human goat. Thus, a carved relief of the *Judenfrau* with her Jewish brood, once to be seen on the tower of a barge in Frankfurt, included the figure of a Jew with two unmistakable goat's horns on his head.

The buck or billy goat, as the Middle Ages knew full well, is the devil's favorite animal and was frequently represented as symbolic of satanic lechery. According to popular legend, the devil created the goat that appears in picture and story as the riding animal of every sort of hobgoblin as well as of witches and sorcerers.[2]

In the witch craze that swept Europe toward the end of the Middle Ages, the devil's most usual disguise was that of a goat, which the devotees worshiped and adored, and it was the animal most commonly offered as a sacrifice. So close was the relation between the goat and the devil that an early fifteenth-century illustration picturing four Jews being led by him represents Satan himself as having goats' horns. Of course, these are most blasphemous and obnoxious notions and are not to be believed at all.

Chelm Stories

The story of the goat and the horse: They had a debate as to who was more useful to man. During the course of the argument the goat mentioned that man used his gut to make strings for such instruments as the violin. The horse said, "Yes, but they have to use my tail to make the bow with which to play the instrument. Moreover, they can only use your gut after you are dead while they can cut my tail and I can go on living."

Be like the horse, who gives of himself while living, to help create beautiful music of life.

Proverbial Sayings

The kids (young scholars) thou hast left behind have grown to be goats (*Berachot* 63a).

Black goat's milk and white goat's milk have but one taste (*Genesis Rabbah* 87:5).

Goose

The goose is not specifically mentioned in the Bible. However, an indirect reference to the Bible is found in the Talmud where we read, "If one sees a goose in a dream, he may hope for wisdom," since it says: "Wisdom crieth aloud in the street" (Proverbs 1:20, *Berachot* 57a). Among the Romans the goose was considered to be a wise bird.

Another reference to the Bible is cited by Feliks, who suggests that *barburim avusim* of 1 Kings 5:3, which is normally translated as "fattened fowl," may in fact be domestic geese.

The goose, *avaz* (of dubious derivation), is often mentioned in the Talmud and, together with the chicken and the duck, was a popular bird in many homes. Thus we learn from Abaye that geese and chickens come under the category of domestic birds, the seizing of which, on a festival day, is not considered a capture and is therefore not forbidden, whereas other birds are forbidden to be captured (*Betzah* 24a).

Like chickens, geese were kept because of their food value, but their eggs were not as popular as those of chickens. According to one authority Babylonian geese are to be considered as water fowl (*Chullin* 56b).

From Rabbi Huna b. Judah we learn that the average value of a goose was one *zuz*, and he adds that when its lungs are dressed with spices it would be worth four *zuz* (*Chullin* 49a).

The rabbis distinguish between domestic and wild geese and offer explicit details about each variety. They are two different species and should not be crossed. Furthermore, the domestic goose has a longer beak than the wild species. The wild goose has its genitals discernible from without, and the other has its genitals within. The wild goose becomes pregnant with only one egg at fecundation, whereas the other becomes pregnant with several eggs (*Baba Kamma* 55a, *Bechorot* 8a).

The goose is well-known for the noise it makes. Thus the rabbis place geese on a par with dogs for guarding a besieged town (*Ketubot* 27a).

Geese were known for their feathers, and the rabbis remark that the fat of a goose wing has medicinal properties, but they looked with disfavor on those who forcibly fed geese as can be seen from the following talmudic story:

Rabbah b. Bar Hana related: "We were once traveling in the desert and saw geese whose feathers fell out on account of their fatness, and streams of fat flowed under them. I said to them: 'Shall we have a share of your flesh in the world to come?' (the feast provided for the righteous). . . . When I came before Rabbi Eleazar he said unto me: 'Israel will be called to account for the sufferings of these geese' " (*Bava Batra* 73b).

Such unnecessary suffering comes under the category of "*tzar baaleh hayyim*," which is condemned.

The story is told that a Maggid decided to settle in a small town. When he met the Rav of the community he explained his purpose. The Rav was surprised and protested: "But the community pays its Rav a very small salary. How will you make a living?" The Maggid told the following parable: "A goose belonging to a thoughtless owner often suffered from hunger because her master forgot to feed her. One day the man bought a rooster and placed him in the same coop with the goose. The goose was very concerned. "Now I shall surely starve. There are two of us that shall eat of the same portion." "Do not worry," retorted the rooster, "I can crow when I feel hungry and this will be a reminder to our owner, then we shall both be fed."

As far as migration is concerned, if geese sought new mates each year they would never keep on schedule; hence their permanent pairing enables them to save time by performing many of their prenesting rituals in southern latitudes.

Proverbial Saying

The goose stoops as it goes along, but its eyes peer afar (*Megillah* 14b).

Grasshopper

As its name implies, the grasshopper feeds on grass and green herbs and leaps a fair distance, even without the help of its wings. This characteristic of leaping is confirmed by the Bible, as we read, they "have jointed legs above their feet wherewith to leap upon the earth" (Leviticus 11:21). Insects or creeping things are forbidden to be eaten, but grasshoppers are an exception and are permitted as food (Leviticus 11:21). (Compare 2 Chronicles 7:13.)

The grasshopper is a member of the locust family and there are several species, each with its specific name.

The *hagav* is of doubtful derivation. The B. D. B. Hebrew Lexicon suggests that it is probably a wingless insect. The verse "We were in our own eyes as grasshoppers, so we must have been in their eyes" (Numbers 13:33) has almost become a proverbial saying because the grasshopper here metaphorically refers to the inferiority complex to which many people succumb.

On this verse Rashi comments: "We heard them say one to another, 'There are ants in the vineyards that look like human beings' " (*Sotah* 35a). Rashi seems to compare grasshoppers to ants. Both can be tiny creatures and from a distance look like human beings (see the section on ants in which they are likened to human beings).

The insignificance of people who are compared to grasshoppers is referred to in Isaiah 40:22: "It is he that sitteth above the circle of the earth and the inhabitants thereof are as grasshoppers."

In Ecclesiastes 12:5 is written, "The grasshopper shall drag itself along." Moshe Hayim Luzzatto, basing himself on *Shabbat* 152a, compares *hagav* with *agav*, "to make love" (compare Ezekiel 23:11, 33:32). Hence passion becomes a burden or drags itself along.

The large variety of *hagavim*, grasshoppers, is attested to by the Talmud where one teacher observes that there are as many as eight hundred (*Chullin* 63b).

Grasshoppers do not multiply at the same rate as locusts, but they can become a plague (*Bava Metzia* 9:6). There is a reference to grasshoppers or locust horns in *Pesachim* 48b, while in *Mishnah Shabbat* 9:7 they are used as children's playthings.

Another type of grasshopper, and a permitted food (Leviticus 11:22), is *saleam*. The Jewish Publication Society edition of the Bible translates this word as "bald locust." The B. D. B. Hebrew Lexicon translates it as swallower or consumer.

The Talmud distinguishes the *saleam* from other types of locusts. It has an arched back in contrast with the *hagav*, but like the *hagav* it has no tail (*Chullin* 65a). Its antennae are very thin. The spring of water oozing out of the Holy of Holies is of the thinness of the *saleam's* antennae (*Yoma* 77b). In one talmudic passage the *saleam* is identified with the long-headed grasshopper, which is named *ail kamtza* or *shushiba* (*Avodah Zarah* 37a).

Yet another kind of grasshopper is the *chargol* mentioned only once in the Torah (Leviticus 11:22), and it is permitted as food. The B. D. B. Hebrew Lexicon calls the word a *quadril* and traces it to a root meaning to run right and left. Jastrow derives it from the root *hegel*, "to go around," with a *reish* inserted.

The rabbis call it *nippol*, probably from the root, *naphal*, "to fall"; they spring and fall on the ground. Another name is *rishon* (*Chullin* 66a) from *rosh*, "head."

The *chargol* is distinguished from the locust because it has a tail, while the locust has none. It also has a protuberance on the back while the *arbeh* (locust) has none (*Chullin* 65b). In the Talmud there is a reference to the *chargol's* egg being carried in the ear for earache (*Shabbat* 67a).

Proverbial Saying

We were in our own sight as grasshoppers (Numbers 13:33).

Hare

This animal is mentioned in Leviticus 11:6 and Deuteronomy 14:7 and is unclean because it possesses only one of the distinguishing marks of *kashrut*, namely, it chews its cud. This may be due to the constant moving of its jaw, which gives the impression that it chewed its cud. Modern authorities believe it is not a ruminant, although the Talmud does describe it as one (*Chullin* 59a).

The Hebrew name for hare, *arnevet*, according to Rav is probably derived from the root *anev*, meaning "to spring"; therefore *arnevet* would convey the meaning of springer. Indeed, the hare is known for its speed in escaping from those who attempt to ensnare it.

An interesting reference to the hare is found in the Talmud, where we learn that King Ptolemy brought together seventy-two elders to transcribe the Torah, and all were prompted by God to conceive the same idea. They wrote for him "the beast with small legs"; and they did not write "the hare," because the name of Ptolemy's wife was hare, lest he should say, "The Jews have jibed at me and put the name of my wife in the Torah" (*Megillah* 9b). In point of fact it was Ptolemy's father who was called hare. From the above we see that *arnevet* is traced from the Greek *dasypoda*, "hairy-footed." The fur of the hare was used for weaving (*Menachot* 39b).

The Almighty assists, in various ways, every creature to preserve itself against attack. Thus, the hare is of a reddish brown color in the summer, but in the winter it turns white so that it cannot be distinguished from the snow. It also has long ears that move in every direction; thus it hears the slightest sound of any potential enemy. Its eyes can see on all sides, and its hind legs are long with strong muscles that help it to spring and jump onto high ground and so elude its pursuers. In addition, it takes a zigzag course in its movements, hides in bushes, and even leaps into the water to swim to safety. In the English translation of the Septuagint this animal is mistakenly taken for the hedgehog.

Proverbial Saying

The hare is an allusion to Yavon (Greece) (*Megillah* 9b).

Hart and Hind

The hart is the male of the herd, and the hind, or the deer, is the female.

In Hebrew, *ayal* (not to be confused with *ayil*, a ram) is a generic term for hart, stag, and deer. The derivation of *ayal* is uncertain, but some connect it with a word meaning "strength or help." Compare Psalm 88:5 where the word *eyal* means "strength or help."

The place named Ayalon (deer field) is connected with *ayal* and is found in Joshua 10:12 and 19:42.

The hart is a clean animal, permitted to be eaten (Deuteronomy 12:15). Indeed it was served at the table of King Solomon (1 Kings 5:3). It is characterized by its elegance and speed; according to the Talmud it is the swiftest animal (*Ketubot* 112a).

In Genesis 49:21 Naphtali is compared to a hind let loose, an image of swiftness. The comment of the Midrash refers to the rapid ripening of fruit. As the Rabbis remark: This is symbolic of the Valley of Gennesareth, which ripens its fruit very quickly, just as the hind runs rapidly (*Genesis Rabbah* 99:12).

The swiftness of the hart is employed metaphorically by the prophet Isaiah, who foretells that even the lame shall leap as a hart (Isaiah 35:6).

King David, too, praises the Almighty for making it possible for him to flee—"who makes my feet like hinds"—as he ascends the heights to escape from his enemies (Psalm 18:34). In an amazing outburst of religious fervor David exclaims, "As the hart pants after the water brooks, so pants my soul after Thee, O God" (Psalm 42:2). Here, too, David uses the simile of the hart to express the relentless flight from his pursuers, who deprive him of the living waters of the Torah for which he thirsts.

The Yalkut analyzes the words of Psalm 42:2 in a striking manner: As the hind gasps for breath and water in her travail, she finds it difficult to conceive; she prays to the Almighty, who answers her prayer by sending a snake who opens the orifice of the matrix and so facilitates birth. So the sons of Korach when (in the superscription to the psalm) they pray to God in their trouble, and He answers them (*Yalkut Shimoni* 741).

On the words "as the hart pants after the water brooks" the Yalkut points out that the hind is the most pious of all animals. For when the animals thirst for water they gather round the hind who digs her horns deep into the earth and prays for water, and the Almighty in his abundant mercy causes the water to rise from the depths of the earth, quenching the thirst of the animals.

Psalm 22 is given the beautiful title *Ayelet HaShachar*, "the hind of the morning."

Modern commentators claim that this title was the name of a melody to the accompaniment of which the psalm was written. The rabbis, however, are more specific. They declare that the antlers of the hind branch off this way and that way, so the light of the dawn is scattered in all directions (*Yalkut Shimoni*).

An interesting observation is made by the Rabbis who infer that Queen Esther recited Psalm 22 when she presented herself to King Ahasuerus (*Megillah* 15b). Rabbi Zera asked why was Esther compared to a hind? To teach you that just as a hind has a narrow womb and is desirable to her mate at all times as at first, so was Esther precious to King Ahasuerus at all times as at the first time. Rabbi Eleazar said, why is the prayer of the righteous compared to a hind? To teach you that as with the hind, it grows its antlers from additional branches every year, so with the righteous, the longer they abide in prayer, the more will their prayer be heard (*Yoma* 29a).

In another biblical passage we read of the beauty of form and gracefulness of the hind: "a lovely hind and a graceful doe" (Proverbs 5:19).

Another name for a male deer is the stag. This animal is found in the Midrash in a striking parable. The rabbis exhort us to love the proselyte or stranger who is compared to a stag among the flock.

The Holy One loves the proselytes; to what is the matter like? To a king who had a number of sheep and goats that went forth every morning to the pasture and returned in the evening to the stable. One day a stag joined the flock and grazed with the sheep, returning with them. The shepherd said to the king: "There is a stag that goes forth with the flock and grazes with them and comes home with them." The king then loved the stag and commanded that no one should beat it.

He ordered that the stag should have plenty of food and drink. The shepherd complained to the king that no special directions were given regarding the sheep, goats, and kids, but only to the stag. The king in his reply pointed out that it is usual for sheep to graze together on cultivated ground, but the stag is found in the wilderness. We must therefore show special consideration for the stag to leave the wilderness and insist on joining us and mixing with the sheep and goats. Similarly, The Holy One enjoins us to love the stranger who leaves his family and background and comes to dwell with us, adopting our customs and laws (*Numbers Rabbah* 8:2).

The stag also figures prominently in a remarkable passage dealing with the teaching of Torah. Rabbi Hiyya said, "I bring flax seed, sow it, and weave nets from the plant. With these I hunt stags with whose flesh I feed orphans, and from whose skins I prepare scrolls, and then proceed to a town where there are no teachers of young children, and write out the five books of the Pentateuch for five children respectively, and teach another six children. . . "(*Ketubot* 103b).

Proverbial Sayings

He placed his money on the horns of a hart (a bad deal) (*Ketubot* 104b).

Why is Jochebed likened to a hind? Because she reared the beauties of Israel (*Chullin* 79b).

Hawk

The Hebrew word *netz* is usually translated as "hawk" and is a generic name including hawk and falcon. It is derived from the root *natzatz*, which is of uncertain meaning.

The hawk is an unclean bird and is mentioned in Leviticus 11:16 and Deuteronomy 14:15. It is swift in flight and rises high in the air in a spiral manner. It belongs to the falcon tribe and there are a variety of species.

The hawk is a migratory bird, as is evident from Job 39:26: "Does the hawk soar by your wisdom and stretch its wings toward the south?" Ibn Ezra expressly states that the hawk migrates to the south for warmth. It is interesting to record that today one can witness tens of thousands of birds migrating at the same time in Israel, as in the period of the Bible.

An interesting detail about the hawk is found in the *Sifra Shemini* 3:5. In Leviticus 11:16 the words "after its kind" refers to the *daya*, a bird of the hawk species.

The keen-sightedness of the hawk is emphasized in *Leviticus Rabbah* 15:2. Rabbi Levi said: The *daya* (mentioned in *Chullin* 63b) is a species of hawk, which, according to Rabbi Johanan, sees its food at a distance of eighteen miles.

We learn from the Mishnah (*Chullin* 3:1) that the *netz* attacks small birds, "small birds that have been mauled by a hawk."

In the period of Edward III (mid-fourteenth century) it was a felony to kill a hawk and to steal its eggs, and the punishment was imprisonment for a year.

In the English language a hawk is used figuratively to mean an officer of the law who pounces on criminals.

Heifer

The heifer plays a significant role in the law known as *Eglah Arufah*, as ordained in Deuteronomy 21:1. "If one be found slain in the land which the Lord Thy God giveth thee . . . and it is not known who hath smitten him . . . the victim is unidentified and the killer unknown." The Torah then advises that the following ceremony should be adopted. "Your elders and judges shall go out and measure the distances from the corpse to the nearby towns. The elders of the nearest town shall take a heifer which has never been worked. . . . Then all the elders of the town nearest the corpse shall wash their hands over the heifer whose neck has been broken and say 'Our hands did not shed this blood, neither have our eyes seen it. Forgive, O Lord, your people Israel whom you have redeemed and do not let guilt for the blood of the innocent remain among your people Israel' " (Deuteronomy 21:2–8).

This seemingly enigmatic ceremony has received the attention of many of our commentators. Rashi, quoting the rabbis of the Talmud, asks: "What kind of confession is this? Would anyone dare imagine that the elders themselves were suspect of the actual bloodshed? Obviously, the leaders are speaking for the entire community who are responsible for the unresolved crime. The elders should ask themselves if they have done everything in their power to prevent the crime that was perpetrated."

The Jerusalem Talmud interprets the Torah text to exclude the mute, the amputees, and the blind from serving as the town elders. The elders of justice, concludes the Jerusalem Talmud, must be without any disability to do justice; they must be perfect in all their limbs.

Therefore, leaders cannot be excused for acting as if they were mute, powerless, or blind when evil or negligence appeared in their community.

Although some classical medieval commentators see, as do some of the modern-day commentators, in the ritual of the broken-necked heifer a mysterious, even magical, rite of absolvement, others see it mainly as a pragmatic device used to handle an unsolved murder. For example, Maimonides states that the "utility is manifest." In most cases the killer is one of the inhabitants of the nearest town. If his identity were known to some people in town who preferred to keep silent, the awesomeness of the ceremony would surely have a strong psychological effect and would consequently intimidate some people in the community to come forward with the information. "The notion is corroborated," says Maimonides, "by the fact that the place in which the heifer has her neck broken should not ever be tilled or sown. Consequently, the owner of that land will use all strategems to investigate in order to discover the killer, and that thus the heifer will not have her neck broken, and his land will not be prohibited from him forever."

The Abravanel finds yet another practical result of the much-publicized ceremony of the breaking of the heifer's neck. It would surely spread the news far and wide that an unknown person had been found slain. This would help free the slain person's wife who otherwise might be left an *agunah*, a woman abandoned by her husband without a divorce, making it impossible for her to ever remarry, due to lack of evidence of his death.

Whatever the reasons for the ceremony, even just reading about it conveys the profound shock of Torah at the possible event of a murder. This utter abhorrence of bloodshed was for all generations a typical characteristic of Jews. Following the news in our own day makes one wonder.

Proverbial Sayings

Egypt is like a very fair heifer (Jeremiah 46:20).

Why was she called *Eglah*, because she was beloved by him as an *eglah* (calf) (*Sanhedrin* 21a).

Heron

The derivation of the Hebrew word *anafah* (heron) is unknown. Some conjecture that it is connected with *anafah*, meaning "face or countenance," as its long beak gives it a formidable countenance. However, the rabbis connect it with the root *anaf*, "to be angry." Why is it called *anafah*? Because it quarrels with its kind (*Chullin* 63a).

The heron is an unclean bird and is mentioned in Leviticus 11:19 and Deuteronomy 14:18. *Anafah* is the generic term for several species of heron; what is common to all of them is a comblike growth on the inner side of their third toe. They feed mainly on fish, and in their insatiable appetite they can devour as many as fifty large fish in one day.

The heron appears in a parable of Rabbi Joshua ben Hananiah in which he tried to pacify Jews who were ready to revolt against the Romans.

A wild lion killed an animal and a bone stuck in his throat. He then made it known that he would reward anyone who could remove it. An Egyptian heron that has a long beak came and pulled it out and demanded its reward. The lion replied: "Go and boast that you entered the lion's mouth in peace, and came out in peace (unscathed)" (*Genesis Rabbah* 64:8).

Hoopoe

The derivation of the Hebrew word *dukifat* is not known, but the Talmud interprets it as *hodo kafoth*, which means its crest was thick, or double (*Chullin* 63a).

In addition to its glorious crescent, or crown, the hoopoe boasts a beautiful plumage that makes it an outstanding bird. However, to offset its natural beauty its nest has a most offensive odor. This characteristic is also providential inasmuch as it deters birds or beasts of prey from attacking it.

The hoopoe is mentioned twice in the Bible (Leviticus 11:19; Deuteronomy 14:18) and is an unclean bird.

The revised version of the Bible translates the name as lapwing, but hoopoe is more generally used. It is interesting that the *Targum Onkelos* on Leviticus 11:19 has the words *nagar tura*, literally, the "carpenter of the mountain," a name for the woodcock. The Talmud explains that this bird guards the *shamir*, the legendary worm that cut the massive stones used for the building of the Temple. Thus, the hoopoe is called *nagar tura*, "cutting through the mountain" (*Gittin* 68a–b).

In the words of the Talmud: What does the bird do with the *shamir*? He takes it to a mountain where there is no cultivation and puts it at the edge of a rock that thereupon splits, and he takes seeds from trees, brings them, throws them into the opening, and things grow there. (This is what the Targum means by *nagar tura*.)

They sought out a woodpecker's nest with young in it and covered it over with white glass. When the bird came it wanted to get in but could not, so it went and brought the *shamir* and placed it on the glass (*Gittin* 68a–b).

A fascinating story of Simeon, who was a keen observer of nature, is recorded in the Midrash: Rabbi Simeon had an orchard, and on one occasion he was sitting in it when he saw a hoopoe was constructing for itself a nest in the trunk of a tree. The rabbi thought to himself, what does that unclean bird want in the orchard? He went and broke the nest, and the hoopoe promptly repaired it. The rabbi then brought a board and placed it over the face of the nest and put a nail through it. What did the hoopoe do? It brought a certain herb, placed it on the nail, and burned it (according to another version "drew it"). Rabbi Simeon then decided to hide the herb so that thieves would not learn to do likewise and ruin mankind (*Leviticus Rabbah* 22:4, *Ecclesiastes Rabbah* 5:5).

The Turks call the hoopoe a messenger bird, and, in one version of a legend relating to Solomon and the Queen of Sheba it is the lapwing or hoopoe that delivers the letter sent by King Solomon inviting the Queen of Sheba to visit him. The letter is found in the *Targum Yerushalmi* on the Book of Esther, chapter 1.

Solomon thought he would make for himself a magic carpet that was invisible and yet would be able to move in any direction at word of command. One day he told Genu, who attended him, that he would fly southward. He commanded: "Carpet, become small enough to hold me in comfort and fly with me southward. Do that I may see manifold variations of life that God has made."

Immediately Genu took up the carpet and the king was wafted southward. As he traveled, the sun became hotter and hotter. When the sun was directly overhead, the king tried to shelter himself from its rays. He saw some vultures flying near him. He called, "Come spread your wings, fly over my head and shelter me from the sun." But the vultures refused–the sun helps us to find decaying food on which we live; we will not offend the sun. The sun grew hotter and hotter and a small cloud appeared and the king could see presently a flock of birds flying toward him. He called out to them to shelter him from the sun's rays and they answered, "Willingly, O king—we are very small but we are numerous, if we spread our wings, they may be a protection to your head." In gratitude King Solomon promised them a reward.

He said, "Come and tell me in three days' time what you would like." For three days the birds set up a council but could not come to an agreement until the king of hoopoes suggested that their queen should choose a suitable reward. She answered, "Ask that all hoopoes be decorated with gold crowns on their heads." All present agreed that this would be a suitable reward, and they flew to King Solomon and asked for the golden crown. Solomon was shocked and exclaimed, "Vanity of vanities—I warn you that admiration often creates jealousy, but if that is your wish, you shall have it." Trouble first came with the fowler and his gun, as he killed many of them for their beauty. Then the hunters set traps of cages with looking glasses inside. The foolish hoopoes came to see how beautiful they had become, but they found they had been made prisoners. Then they were sold for their plumage. In great alarm the king of the hoopoes went to Solomon: "What have you done to us; you have returned evil for good. Before we helped you, none sought our lives, now we are in danger of being utterly destroyed." King Solomon replied: "I now see that some creatures are incapable of choosing best for themselves and it is necessary for the wiser men rulers to help them choose. I warned you that vanity would be your downfall. Now let me choose. I suggest that all golden crowns be changed to feathers." And turning the magic ring the king pronounced the necessary words and it was done. The hoopoes said, "Wise and great is Solomon the king—he has given us warmth and security."

Hornet

The derivation of the Hebrew word *tsirah* is not certain, but Ibn Ezra connects it with *tzarat*, meaning "leprosy." This is based on the supposition that the poison of the hornet causes an infection on the skin that resembles leprosy.

The hornet is a ferocious outsized wasp, and it is mentioned three times in the Bible: Exodus 23:28, Deuteronomy 7:20, and Joshua 24:12.

Rashi, on the passage in Exodus, quotes the Talmud, which states that the hornet did not pass over the Jordan with them but stood by the bank and injected a virus into the Canaanites that blinded their eyes above and castrated them below (*Sotah* 36a).

Interpreting Deuteronomy 7:20, one commentator suggests that the hornets alone could not overcome an army.[1] He therefore is of the opinion that the hornet was symbolically used by some of the Pharaohs in their invasions, thus reducing the fighting power of the Canaanites. However, the destructive nature of the hornet's sting is attested to by Driver, who writes that if a horse treads upon a hornet's nest, "it would have to fly with all speed, for the combined attack of a swarm of hornets has been known to be fatal."[2] Indeed, the evil effects of a wasp sting are recorded in the Talmud where we learn that a wasp came out of a wall and stung a Galilean on the forehead and he died (*Shabbat* 80b).

In another passage the rabbis maintain that it is permitted to kill the hornet of Nineveh on the Sabbath. However, one teacher, Rabbi Huna, objected, and he chided a man for killing a wasp on the Sabbath (*Shabbat* 121b). Incidentally, the hornet of Nineveh is identified with the bee of Assyria (Isaiah 7:18).

The Oriental variety of the hornet is found in all parts of Israel.[3] In common with the bee it lives in societies where there is division of labor among the queen, workers, and drones. It likewise builds a comb with characteristic hexagonal cells, though not of wax, but a paperlike substance. The honey of hornets is permitted for eating (*Machshirim* 6:4).

The Midrash mentions the hornet as a punishment used by God against man. In all places the Holy One carries out His purpose, and He has not created a single thing in vain. Sometimes He achieves His purpose by means of a frog, hornet, or scorpion. God inflicts punishment by means of tiny creatures to teach us that the alleged strength of the enemy is not real (*Numbers Rabbah* 18:22).

The expression "stirring up a hornet's nest" has entered the English idiom. It means to stir up a host of virulent enemies around one.

Midrash Tehillim reads: Whenever the Amorites came out of hiding, one hornet would strike in one eye and another hornet in the other eye. (*Yalkut Shimoni* reads: The Amorites would hide in a cave and place a stone at the mouth of the cave, but two hornets would render the stone asunder.)

In the essay on the spider, there is a story about the hornet.

Proverbial Sayings

Say to the hornet: "Neither your honey nor your sting" (*Tanchuma Balak* 6).

"Its honey for its owner and its sting for others" (*Midrash Rabbah*).

Horse

The common Hebrew name for the horse is *sus*, which is often found in the Bible; its derivation is unknown. Other words for horse include *rechesh*, "a swift steed" (Micah 1:13, Esther 8:10, 14), which is translated by the authorized version as "mule," and *Parashim*, found only in the plural, which is rendered as "horsemen" (Genesis 50:9). In Judges 5:22 some translate *abirim* as "magnificent steeds" (compare Jeremiah 8:16).

In early Israel the horse was never included among the domesticated animals, such as the ass or ox. Indeed the ass was a symbol of peace, but the horse was a symbol of war, and in the Bible the horse is always represented, with few exceptions (e.g. Isaiah 28:28) as a fighting animal.

The earliest reference to the horse in the Bible is found in Genesis 47:17: "Joseph gave bread in exchange for horses." From this we learn that horses were commonly found at that time in Egypt. This fact is emphasized in the command to the king not to multiply horses, in order to restrain the people from returning to Egypt (Deuteronomy 17:16). This command obviously refers to the warhorse, for kings measured their military strength by the number of horses they possessed.

In 1 Kings 10:28 the verse clearly states that "the horses which Solomon had were brought out of Egypt." In 1 Kings 5:6 we are informed that Solomon had forty thousand stalls of horses for his chariots, but in 2 Chronicles 9:25 the number is estimated as four thousand stalls. It is possible that the stall in one passage may refer to a stall containing ten apartments, each for one horse, and in the other passage, each apartment is called a stall.

It is worthy of note that excavations at Meggido confirm that Solomon had four thousand horses. A fascinating picture of the layout of Solomon's stables with a stone trough for each animal and the reconstruction of the feeding, lighting, and ventilation are given by Thieberger.[1] Egypt was not alone in possessing horses. Rav Shakeh, the Assyrian commander, contemptuously challenged King Hezekiah by offering him two thousand horses if he could find men in his army to mount them (2 Kings 18:23). In 2 Kings 23:11 King Josiah condemned horse worship to the sun, a form of idolatrous worship that was practiced by Armenians and Persians.

The most vivid portrayal of the horse in the Bible is found in the Book of Job: "The glory of his snorting is terrible. He paweth in the valley, and rejoiceth in his strength: he goeth out to meet the clash of arms. He mocketh at fear, and is not affrighted; neither turneth he back from the sword" (Job 39:20–22).

To counteract the alleged glory and strength of the horse, the Psalmist warns us, "Some trust in chariots and some in horses, but we will make mention of the name of the Lord our God" (Psalm 20:8). Compare Psalm 33:17: "A horse is a vain thing for safety, neither does it afford escape by its great strength."

Before leaving the Bible, a brief mention should be made of "The Horse Gate" of the Temple, which was in the east section of Jerusalem overlooking the Kidron. It opened into the southeast corner of the Temple and courts (see Nehemiah 3:28, 2 Kings 11:16, Jeremiah 31:39, 2 Chronicles 23:15).

In Talmud, and Midrash, as well, the horse plays a prominent role. Some interesting data emerge. Six things were said of a horse:

1. It loves promiscuity (compare Ezekiel 23:20).
2. It loves battle (and is therefore called a warhorse).

3. It is high-spirited.

4. It needs little sleep; it indulges only in 60 respirations[2] (*Sukkah* 26b, *Berachot* 3b).

5. It eats much.

6. It excretes little (*Pesachim* 113b).

The popularity of the horse is seen in this passage: "Do not dwell in a town in which no horses neigh or dogs bark" (*Pesachim* 113a). (These animals guarded the town. The dogs barked, and the bandits were pursued on horseback.)

Both in the Song of Moses (Exodus, chapter 15) and the Song of Deborah (Judges, chapter 5), the horse plays a vital role.

"The horse and his rider He has thrown into the sea" (Exodus 15:1). Both here and in Exodus 14:23 the rabbis note that the singular *sus* is employed to teach us that they were all accounted before God as one horse (*Mechilta*).

Exodus Rabbah 21:5 gives another interpretation: The Lord took first the guardian angel (warhorse) of the Egyptians and drowned him.

In Judges 5:22 there is a realistic picture of the galloping horses that stamp the ground and with the chariots make a terrific noise: "Then were the horse hooves broken by the prancings, the prancings of their mighty ones."

We now turn our attention to the Midrash.

Elisha ben Avuyah said that a man who has learned much Torah and has good deeds is like a horse that has reins. The man who has the first, but not the second, is like a horse without reins; it soon throws the rider over its head (*Avot d'R. Nathan* 24:4).

A similar thought is expounded by the Yalkut on Psalm 32:9: "Be not as the horse or as the mule which have no understanding; whose mouth must be held in with bit and bridle, that they come not near unto you." The Yalkut warns man not to behave as a horse or mule; when you feed it with barley it kicks. . . . Come near to it and it kicks, but you, do not act in a similar manner when God wishes to reward you (*Yalkut Psalms* 719).

The horse is employed in the following parable.

Commenting on Genesis 20:6, "and I withheld you from sinning," the Midrash says this is like the case of a warrior riding his own horse at full speed. When seeing a child lying in his path, he reins the horse so that the child should not be hurt. Do all praise the horse or the rider? Surely, the rider! Similarly, here God said: "I did not allow Avimelech to touch Sarah" (*Genesis Rabbah* 52:9).

In another passage the horse is praised for his preparedness to serve his master in battle. In quoting Proverbs 21:31 Pinchas the priest argues in this manner: "If the horse who receives no reward risks its life in the day of battle and is prepared to die for its master, how much more so should I be prepared to die in order to sanctify God's name" (*Pesachim* 53b).

A story is told of the Gerer Rabbi, who in a discourse on life after death remarked that it is impossible for an unlearned man truly to enjoy Paradise, even though he may honestly deserve it. He illustrated his point by relating a story: A *tzaddik* was traveling in the early spring in a wagon. The roads were in terrible condition. The axles of the wheels broke several times, and the horses plowed with difficulty through the slush and mud. Friday morning came, and a great distance was yet to be covered before the *tzaddik* could reach his destination. He turned to the teamster and said, "It is important that I arrive at my goal before the advent of *Shabbos*." The teamster promised to do his best. A horse fell dead from exhaustion, but the teamster continued with the second horse, and succeeded in reaching the *tzaddik*'s destination before *Shabbos*. On Sunday the *tzaddik* heard that the second horse also died and that the teamster's grief was so great that he became ill. The *tzaddik* ordered the best medical attention for him but in vain; he died.

When his soul came before the heavenly tribunal the counsel for the defense won the case, and Paradise was ordained for the poor teamster. His soul arrived there, but it found no pleasure in the spiritual and cultural environment of even the lowest region. He was then sent into an imaginary world where he was presented with a beautiful carriage harnessed to four magnificent horses and where the roads stretched before him, always dry and always level. The teamster was able to enjoy only an imaginary Paradise, not the true one.

Consider the tale of the horse owner who came to his rebbe to seek advice about his horse, which was behaving in a strange manner. The animal rejected his usual diet of hay and oats and, instead of water, he insisted on the finest beverages. "He also refuses to sleep on the stable floor and insists that I supply him with a mattress and pillow."

The rebbe asked, "Tell me, did you say your prayers this morning?" "I got up too late," answered the *chasid*. "Did you wash your hands before you sat down to eat? Did you say grace after the meal, did you say *Minchah*?" To all he answered, "No." Then the *chasid* asked, "What have my religious failings got to do with my horse's ridiculous actions?"

The rebbe replied, "Can't you see—your horse, noting too that you were behaving as a beast (animal), decided that he would conduct himself like a human being and demand special treatment."

Man must therefore discard his animalistic tendencies.

Proverbial Sayings

Poverty is becoming to a daughter of Jacob as a red strap to a white horse (*Song of Songs Rabbah* 1:24).

Drive your horse with oats, not with a whip (Yiddish proverb).

Abraham, Isaac, and Jacob ran before God as horses (eagerly and swiftly) (*Sanhedrin* 96a).

Horse Leech

The horse leech is a large aquatic sucking worm and is only mentioned once in *Tanach*. "The horse leech has two daughters: Give, Give" (Proverbs 30:15).

The Hebrew word for horse leech, *alukah*, is derived from the root *alak*, meaning "to cleave, to adhere." The leech cleaves to the nostrils and mouth of the person or animal who drinks from water infested with them. The rabbis were fully aware of the dangers of this worm and gave explicit instructions to avoid being attacked. Thus the Talmud advises that one should not place one's mouth on a water pipe and drink from it for fear of danger. What is the danger? The swallowing of a leech. Our rabbis taught: "One should not drink water either from rivers or from a pool directly with his mouth or by drawing the water with the one hand . . . if he drinks it, his blood shall be upon his head . . . for he can swallow a leech." Rabbi Hanina said for one who swallows a leech it is permissible to get water heated on the Sabbath (*Avodah Zarah* 12b). This Talmud continues this discussion and cites a special case; Huma suggested that the person so affected should drink vinegar and not water.

In another instance the rabbis suggest that the bug drives out the leech (*Berachot* 9:2), the underlying reason being that the leech is driven away by the odor of the bug. Jastrow explains that it refers to a potion mixed with bedbugs that will cause the removal of the leech as a result of vomiting. In *Bechorot* 44b a case is reported of a leech causing the swelling of the stomach.

The rabbis offer an allegorical interpretation of the verse, Proverbs 30:15. Mar-Ukba interprets the words "Give, Give" in the following manner: The horse leech has two daughters, "Give, Give." It is the voice of these two who cry from Gehenna, calling to this world "Bring, Bring!" And who are they?—*Minut* (which continuously lures the unwary to its erroneous teaching) and the government (which constantly imposes new taxes and duties) (*Avodah Zarah* 17a).

Another form of *alukah* is *elka*, which is found in the *Targum* on Psalm 12:9 where *zulut*, an unusual word meaning "worthlessness," is paraphrased in the following manner: Return you wicked men who walk like a leech that sucks the blood of men. This is an unusual interpretation of the ways and habits of wicked men.

One writer explains the "two daughters" of Proverbs 30:15 as follows: This reptile appears to have the power of suction at the tail as well as at the head, by means of which it can affix itself firmly to the inside of the glass in which it is sometimes placed. This power of suction at either end might suggest the idea of two daughters.[1] The simple interpretation of the verse is a warning against the insatiable greed that overtakes some people who aggrandize themselves at the expense of the poor and needy whom they oppress and persecute by sucking their lifeblood and enriching themselves with their unjust gain.

Hyena

The hyena is not specifically mentioned in *Tanach*, but it has been suggested that *haAyit tzavua* in Jeremiah 12:9, which is translated as "a speckled bird of prey," might well be the hyena. The Hebrew *tzavua* would then be derived from *tzeva*, "dye or color." Jastrow renders this phrase as "the checkered *tzavua*, leopard, or striped hyena."

The *tzavua* is also the source for several place names. Thus, *Gei Tzevo'im* in 1 Samuel 13:18 might perhaps be rendered as the valley of hyenas (compare Nehemiah 11:34). *Tzevo'im* is also mentioned in *Bikkurim* 1:3. Some connect *tzavua* also with proper names, such as Tzivon in Genesis 36:2, 14, 20 and in 1 Chronicles 1:38, 40.

In a revealing passage in the Talmud we learn that the male *tzavua* (hyena) after seven years turns into an *arpad*, the *arpad* after seven years turns into a *kimmosh*, the *kimmosh* after seven years turns into a thorn, and the thorn after seven years turns into a demon (*Bava Kamma* 16a). Because of these physical changes through which the hyena passes, it is known by a variety of names: *barbalis*, *nafraza*, and *apa* (Bava Kamma 16a).

The Talmud Yerushalmi reminds us that the *tzavua* can be as fierce as a lion (*Bava Kamma* 1:5). In Ecclesiasticus 13:18, we learn that there can never be peace between the hyena and the dog. One is rich and the other is poor.

The hyena feeds on offal and is also known to exhume bodies. Its teeth and jaws are so strong that it can crush the bones of an ox. In the

Talmud we read that King Alexander Janneus in the first century B.C.E. warned his wife not against the true Pharisees, but against the "dyed" Pharisees, who were like hyenas or chameleons, doing the deeds of Zimri but expecting rewards like Pinchas (*Sotah* 22b).

Genesis Rabbah 7:4 states that the *tzavua* is formed from a white drop and has 365 colors.

Proverbial Saying

Like the hyenas doing the deeds of Zimri but expecting rewards like Pinchas (*Sotah* 22b).

Jackal

There is some confusion over the translation in Hebrew of the jackal. Some connect the jackal with *shual* (fox).

The Jewish Publication Society of America translates *ee'im* (Isaiah 13:22) as "jackals shall howl." The B. D. B. Hebrew Lexicon also translates this word as "jackal" (howler). This animal is an inhabitant of desert ruins and is known for its mournful howl (see Isaiah 34:14, Jeremiah 50:39).

Yet, the Hebrew word *tanim* is also translated as "jackal," and again its mournful howling is emphasized. Compare Micah 1:8: "I will make a wailing like the jackals."

Tanim is also used in Job 30:29: "I am become a brother to jackals." Ibn Ezra comments on this verse: "I make a wailing like jackals: They are wild animals that have a cry of desolation." In former days the cry of the jackal was often heard at night, but today it has almost disappeared.

In contrast, the *Targum Jonathan* translates *tanim* as *yarodin*, and the Talmud Yerushalmi reports that "the *yerodot* are to be considered birds in every respect" (*Kilayim* 8:4). *Yarod* is also found in Talmud Bavli, *Sanhedrin* 59b, where it is used in a contemptuous manner to describe a rabbi who is called "a howling *yarod*," which is a bird of solitary habits, or according to Rashi, a jackal.

Again in *Ketubot* 49b we learn from Rav Judah that a *yarod*, the Hebrew *tanim*, "bears progeny and throws them upon the tender mercies of the townspeople." Here again, Rashi translates *tanim* as "a dragon or jackal" (compare Jeremiah 9:10).

According to the Jewish Encyclopedia (12:188), in ancient times the tombs were closed by means of large stones in order to protect them against the ravenous jackals.

Leopard

The leopard is known in Hebrew as *namer*, (plural, *namerim*). It is mentioned several times in the Bible, and its derivation is uncertain. The B. D. B. Hebrew Lexicon connects *namer* with the Assyrian root meaning "to shine, gleam," as does the coat of the panther or leopard. Jastrow connects *namer* with the Hebrew root *namer*, "to give a checkered or striped appearance." (Compare *Pe'ah* 3:1, "making the field look checkered like a tiger or leopard.")

From the above it would appear that the leopard resembles both the panther and the tiger; its yellow skin has a large number of black spots in circular groups giving the appearance of checker work.

The most oft-quoted passage in Scripture regarding the leopard is in Jeremiah 13:23: "Can the Ethiopian change his skin or the leopard his spots?" As the leopard was characterized by its spots, which were natural and unchangeable, so unfortunately, do evil habits (black spots) cling to a person and can be eradicated only by resolute action and determined will power.

From Habakkuk 1:8 and Hosea 13:7 we learn that the leopard was a swift-moving and agile animal that waited patiently to pounce on its prey.

In Daniel 7:6, a leopard appears to Daniel in a dream, with four wings and four heads. Saadiah Gaon interprets the verse to refer symbolically to the Grecian empire. As the leopard is a ferocious and dangerous animal, so Alexander the Great threatened to subdue the world with the swiftness and rapidity of a leopard and partition the empire among his four captains.

Modern commentators see the leopard as referring to Persia extending its power to the four directions of the compass.

In Isaiah 11:6 we have the glorious messianic picture of the golden era when "the wolf shall dwell with the lamb and the leopard shall lie down with the kid . . . and a little child shall lead them." However, leopards attack human beings as well as beasts: "A leopard watches over their cities, everyone that goes out thence is torn in pieces" (Jeremiah 5:6); "and they dwell in the mountains; from the mountains of the leopards" (Song of Songs 4:8).

In the biblical period, leopards must have been common because of the place named "Nimrah" in Numbers 32:3, and "Bet-Nimrah" in Joshua 13:27, but today the leopard is scarce though it still survives in Israel. The leopard is one of five animals which are considered a danger to the public (*Mishnah Bava Kamma* 1:4).

In addition to its Hebrew name *namer*, the leopard is also referred to by its Greek name, *bardalis*, which Jastrow renders as "a spotted beast." This leads us to a quaint passage in the Talmud: What is *bardalis*? Rav Judah says *nafraza*. What is a *nafraza*? Rabbi Joseph said *apa* (contraction of *apa*, "hyena"). Rabbi Meir adds the *zavua* (many-colored, another term for hyena) because of its colored stripes. Now Rabbi Joseph said that *zavua* means *apa*. This, however, is no objection, for the latter appellation, *zavua*, refers to the male, whereas the former, *bardalis*, refers to the female (Bava Kamma 16a).

In a homiletical interpretation of Nehemiah 7:61 God observes: "I said that Israel should be as precious to me as the cherub whereas they made themselves like the leopard, which is not particular to copulate with its mate, so Israel thereby produces *mamzerim*." According to others Rabbi Abbahu stated: "Though they have made themselves like the leopard yet they are as precious to me as a cherub" (*Kiddushin* 70a).

The Rabbis interpret the verse in Jeremiah 13:23 in a different vein from the standard translation. They suggest that *namer* (leopard) is derived from *mur* (to change), and the verse should therefore be translated as follows: "Can the Ethiopian change his skin or turn (that is, heal) his wounds?" In this manner *habarbarothaw*, normally "spots," would be derived from *haburah*, "a wound" (*Shabbat* 107b).

Finally, the leopard is mentioned in the popular Mishnah of *Avot* and repeated in the first chapter of the *Shulchan Aruch*: Be bold (ferocious) as a leopard . . . to do the will of thy Father who is in Heaven (*Avot* 5:23). Some interpret this Mishnah to point to the pursuit of the study of Torah. "If a leopard, a creature of no intelligence, uses all his daring to seize his prey to get food, how much more should man, a creature of intelligence, use all his daring to acquire life for his soul in the world to come."[1]

The above Mishnah is interpreted in another context. It is written: "Do not glorify yourself in the presence of kings and stand not in the place of great men" (Proverbs 25:6). "If a man is to bear himself humbly before a king of flesh and blood, how much more before God. They teach, be bold as a leopard, and so on, in order to instruct you that there should be no pride before God" (*Numbers Rabbah* 4:21).

Proverbial Saying

Be bold as a leopard . . . to do the will of thy father (*Avot* 5:23).

Leviathan

The word *leviathan* is the transliteration of the Hebrew word that is derived from a root meaning "to coil or twist." In poetic imagery the leviathan, which is a gigantic sea monster, encircles or coils itself around the seas of the world.

The leviathan is known by a variety of names. In Job 40:25, leviathan refers to the crocodile and behemoth; in verse 15, the hippopotamus; and in Isaiah 27:1, the fleeing serpent and coiled serpent, which when coupled with the dragon in the sea, are symbolic designations of world powers. It is also referred to by some commentators as the whale. The Hebrew poets, especially, were fond of using these creatures to symbolize the destruction of Israel's enemies. As to the relation of the Hebrew to the Babylonian conception of the cosmogeny, the position has been most clearly summed up by Driver, who says: "The narrative of Genesis 1 comes at the end of a long process of gradual elimination of heathen elements and of gradual assimilation to the purer teachings of Israelite theology carried on under the spiritual influences of the religion of Israel."[1]

In rabbinic literature it is considered to be a legendary sea animal reserved, with the behemoth, for the righteous in the hereafter (*Leviticus Rabbah* 13:3, *Avodah Zarah* 3b, *Mo'ed Katan* 25b).

The leviathan was created on the fifth day (*Yalkut Shimoni*, Genesis 12). According to *Bava Batra* 74b the leviathan was created in a male and a female form, but there was the fear that it might multiply to such an extent that it would destroy the world. God therefore slew the female of the species.

Many *haggadot* concerning the leviathan are in the name of Rabbi Johanan (see *Bava Batra* 74a–b).

However, in spite of its huge size, the leviathan is afraid of a small worm called "*kilbit*," which clings to the gills of large fishes and kills them (*Shabbat* 77b).

The leviathan is also mentioned in our festival liturgy, for when we leave the *sukkah* at the conclusion of the Festival of Tabernacles it is customary to recite the following short prayer: "May it be Thy will, O God, that just as I have dwelt in the *sukkah*, so shall I be privileged during the coming year to dwell in the *sukkah* made of the skin of the leviathan."

The source of the prayer is found in the Talmud where we read that Rabbah, in the name of Rabbi Johanan, stated: "The Holy One, blessed be He, will in time to come make a tabernacle for the righteous from the skin of the leviathan" (*Bava Batra* 75a). This passage requires clarification.

The leviathan symbolizes the evil forces of violence, persecution, aggression, and war, all of which have menaced the Jew in his long and checkered history. However, in the Messianic Era the leviathan will be overcome, and with its destruction the wicked enemies of Israel will be subdued and anti-Semitism eradicated. In this manner, the daily prayer that we recite in the *Maariv* (evening) service–"Spread over us the tabernacle of Thy peace"–will be fully realized and an era of absolute peace will be ushered in, and the tabernacle will receive a new covering consisting of the skin of the leviathan. In this international *Sukkah* of Peace the righteous will be regaled with the fish of leviathan.

Proverbial Saying

If leviathan by hook be hauled to land, what hope have fishes of a shallow strand? (*Mo'ed Katan* 25b).

Lice

Lice—*ken* or *kinah* (plural, *kinnim*), of doubtful derivation—are mentioned in connection with the third plague that God brought upon the Egyptians (Exodus 8; Psalm 105:31). This plague was sent on the land to smite the dust of the earth.

Alternative translations of *kinnim* as "gnats or mosquitoes" have been suggested, but these insects hatch in water and not in the dust of the earth.

The magicians of Egypt were unable to produce lice through their magic arts and therefore exclaimed, "This is the finger of God" (Exodus 8:15).

It is interesting that the "finger of God" is referred to on two other occasions. In connection with the writing of the Ten Commandments (Exodus 31:18) we read "the tables of stone written with the finger of God." And the Psalmist praises God with these words, "When I behold Thy heavens, the work of Thy fingers, the moon and the stars which Thou hast established" (Psalm 8:4).

Understandably, there was justification for employing the expression "the finger of God" in reference to the Ten Commandments and the heavenly spheres because they speak of the lofty heights of Divine Providence. However, the plague of lice is mundane, and how can we reconcile it with the celestial heights? We should remember that God is not ensconced in the heavens above, but He also reigns over the earth and He created the vermin too.

This thought is echoed by the Rabbis who confirm that God feeds the whole world, from the horned buffalo (the largest) to the eggs of the louse (the smallest) (*Avodah Zarah* 3b).

Proverbial Saying

He who killed vermin on the Sabbath is as though he killed a camel on the Sabbath (*Shabbat* 107b).

Lion

The lion, which is mentioned more than one hundred times in the Bible, is designated by six names:

1. *Aryeh* or *ari*, which according to Jastrow is derived from the root *arah*, "to be light," and is therefore translated as "the light-colored lion." *Ari* may also refer to a distinguished scholar.

2. *Shachal* is a fierce lion of middle age. This word is traced to an Assyrian root meaning "to cry" and may therefore refer to the roar or cry of the lion when he is hungry.

3. *Shachatz* is a lion more advanced in age than the previous one. In the words of the rabbis the lion is called *schachatz* because it tears with its mouth (*Avot D'Rabbi Nathan* [second version], chapter 43).

4. *Lavie*, which Jastrow derives from the root *lavan*, "to be bright, or white," and renders as the flame-colored lion.

5. *Layish*. This is derived from the root "to be strong, lionlike" (B. D. B. Hebrew Lexicon). The *layish*, too, is an old lion.

6. *Kefir* is a young lion, *gur*–"whelp." The etymology is not known but perhaps is onomatopoeic. *Kefir* differs from *gur* inasmuch as it is old enough to hunt its prey (B. D. B. Hebrew Lexicon).

Five of the six names are found in Job 4:10–11.

Today the lion is extinct in Israel, but from the days of the Bible until the Middle Ages it was common. In ancient Palestine the jungle was not far from civilized life, so that the cry "A lion is in the way" was not unusual. Samson killed a young lion that was prowling in the vineyards (Judges 14:5–6), David smote a lion (1 Samuel 17:36), and Benaiah slew a lion in a pit of snow (2 Samuel 23:20).

Animals are different from man; they cannot provide for their future. Thus as the lion becomes too old to catch its prey, it dies of hunger: "The old lion perishes for lack of prey" (Job 4:11).

When the prophet Amos cries out, "Will the lion roar in the forest when he has no prey?" (Amos 3:4), he is issuing a warning of danger to Israel that the lion, the might of Assyria, is threatening to engulf the world (compare Nachum 2:13 and Isaiah 5:29). When the prophet Jeremiah warns that "a lion is gone up from his thicket and a destroyer of nations is set out" (Jeremiah 4:7), he is referring to Nebuchadnezzar of Babylonia.

Again, when Amos asks, "Will a young lion give forth his voice out of his den if he has taken nothing?" the prophet is thinking of Israel's neighbors who are ready to take advantage of Israel's weakness. Finally, we are told that God's messages to his people are conveyed through his prophets–"The lion has roared–the Lord God has spoken, who can but prophesy?" (Amos 3:8).

Perhaps the most popular metaphor regarding the lion is found in Genesis 49:9 in which Judah is a lion's whelp. As the lion triumphs over his enemies, so will Judah. In the words of Samson Raphael Hirsch, "Not in the fighting and in the thick of the fray does Judah's greatness lie . . . even when he is quietly resting he remains a lion." The prophecy of the patriarch Jacob was realized, for from the tribe of Judah kings were raised.

The Lion of Judah, the emblem of the tribe, has left a deep imprint on Jewish life. Lions figured largely in the decoration of Solomon's Temple (1 Kings 7:29; 10:19–20), and the Lion of Judah is found today in almost every synagogue. It is embroidered on the curtain of the Ark and on the mantle covering the Scroll of the Law, is engraved on the silver ornaments hanging on the mantle, and is carved on the reading desk.

In midrashic literature, too, the lion figures in metaphor, simile, and parable. Commenting on Numbers 23:24, "It is a people which rises up as a lion," the Midrash observes: There is no people that resembles them. In sleep they are unconcious of the Torah and the commandments, but they rise up from sleep like lions and hurry to the reading of the *Shema*, and proclaim the sovereignty of God, and strengthened by their prayers, like lions, they separate each to his occupation and business, and if one stumbles, or temptation assails him, he proclaims the sovereignty of God; he does not "lie down till he eats of the prey" (*Numbers Rabbah* 20:19).

In another passage Rabbi Joshua ben Hananiah successfully quelled a revolt by the masses against Hadrian by telling them a fable about a lion who had a bone caught in his throat and who offered a reward to anyone who would extract it. The Egyptian heron appeared and put his long bill into the lion's throat and so relieved him. When he asked for his reward the lion replied that he had already received his reward by escaping unscathed from the lion's mouth. "Similarly," said Rabbi Joshua, "we can now be happy in knowing that Hadrian is allowing us to live undisturbed" (*Genesis Rabbah* 64:8).

In another passage the rabbis spell out the evils of drunkenness. If a man drinks properly, he becomes strong as a lion that nothing in the world can withstand. When he drinks more than is proper he becomes like a pig that wallows in mire, and when he becomes drunk he dances like an ape and utters folly before all (*Tanchuma Noah* 13).

Many are the references to the lion in the Talmud, but this discussion focuses on the popular Mishnah in *Avot*: "Be as strong as the lion to do the will of your Father who is in heaven" (*Avot* 5:23).

Initially, this teaching seems rather strange. We compare the lion who can be cruel and ferocious with the will of God who is kindly, merciful, and gracious. However, the word *gibor* does not refer to the exceptional physical strength of the lion but to the ability with which God has endowed him to curb his passions. This is in the spirit of the teachings of Ben Zoma. "Who is mighty (*gibor*)?" "He who subdues his passions" (*Avot* 4:1).

The lion will devour and plunder when he is hungry and will provide for the needs of the members of his family, but he is not naturally cruel.

In contrast to the bear, the lion will not recklessly hunt for prey. The lion has been known to aid weaker animals and even procure food for them. Indeed, the lion has the strength to subdue his passions. This may explain why the righteous and upright of Israel "are stronger than lions to do the will of their Master and the desire of their Rock" (Sabbath morning prayer, *Av HaRachamim*). It is also reasonable to suggest that Daniel and his companions left the den unscathed because the lions, under Divine Providence, controlled their passions (Daniel, chapter 6).

This evaluation of the "might" of the lion is borne out by our sages in the Talmud where we learn that sometimes the lion will stay among the flocks and not injure them (*Chullin* 53a). We are also informed that the lion will attack man only when it is driven by hunger, but never when two men are together (*Shabbat* 151b). It is in this spirit that the rabbis have called the lion the king of the beasts (*Chagigah* 13b). He has earned this title because he can govern his passions. In this respect some authorities identify the thick shaggy mane round the head of the lion as its crown.

A quaint story is recorded in the Talmud about the lion. R. Simeon ben Halafta was walking on the road when lions met him and roared at him. Thereupon he quoted the verse "The young lions roar after their prey and seek their food from God" (Psalms 104:21). Immediately, two lumps of flesh descended from heaven, and they ate one and left the other (*Sanhedrin* 59b).

We see that even in the narrative above, the lions curbed their appetites and subdued their passions. Commenting on Amos 5:19: "As if a man did flee from a lion, and a bear met him," *Pirkei d'R. Eliezer* declares that the lion means Laban, who pursued Jacob like a lion so as to destroy his life. The bear refers to Esau, who stood by the way like a bear. The lion is shamefaced, the bear is not. Not only did Laban refrain from molesting Jacob, but he admitted that he was unable to do so. However, Esau made no such admission. This fits in with our contention that the lion can control his passion.

Little wonder that the Jew has survived. He will accept the restraints of the moral law and uphold family purity, he will curb his appetite and adhere to the Jewish dietary laws, and in many areas of life he will repress his passions and rightly earn the title, the Lion of Judah.

The lion is also reflected in the *Halachah* that enumerates four types of proselytes, one of them being the lion-proselyte. This is based upon the text in 2 Kings 17:24–25 where we read that the Samaritans were brought by the king of Assyria to Samaria: "They took possession of Samaria and dwelt in towns. When they first settled there they did not worship the Lord; so the Lord sent lions against them: Because they converted due to fear of divine visitation (i.e. lions) and not because of the love of God, they are called 'lion-proselytes.' " Some rabbis suggest that the Samaritans are lion-proselytes, and therefore they are considered as Gentiles (*Chullin* 3b, *Yevamot* 24b).

Commenting on Proverbs 18:21, "death and life in the power of the tongue," the Midrash relates the story of a Persian king who became ill. His doctor said he might get better if he drank milk of a lioness. One of his servants volunteered to try to procure some, taking with him some sheep with which to lure the beast. He succeeded in getting some milk from a lioness. On his way home he slept and dreamt. In his dream different parts of his body disputed as to which of them had done most to contribute to the success of his mission. Feet claimed they did most: "Without us lioness could never have been reached." Hands said, "What was the use of going there if we had not done the necessary work of milking the lioness?" Eyes said, "Neither of you would have been of any use had we not supplied sight to enable the right way to be found. Without us nothing could be done." Heart said, "It was I that conceived the scheme," and was silenced by all the rest who suggested that the tongue imprisoned in the mouth could not possibly have contributed anything to the success of the enterprise. In the middle of the argument, the man woke up. He continued his journey homeward and was led into the king's presence with his precious flask. "Here your Majesty, I have brought you the dog's milk." The king thought that the man was deliberately insulting him and ordered him to be put to death. All the parts of his body were terribly afraid, and the tongue compelled them to acknowledge that its power was greater than that of feet, hands, eyes, and heart. The tongue eloquently pleaded with the king, explaining that it was only by a slip that the milk was described as dog's milk. It was really the milk of the lioness that was procured at great risk. Let the king put it to the test. He did so and recovered from his illness. The man was set free and rewarded. All the members of the body acknowledged that the tongue had spoken the truth (*Yalkut Shimoni, Midrash Tehillim* 721).

The Besht was once asked, "Why does the Bible relate the wrongdoings of good men; would it not encourage righteousness to teach that good men are invariably good?" The Besht replied that if the Bible failed to indicate the few sins committed by its heroes, we might doubt their goodness and he explained this by the following fable: A lion taught his cubs that they need fear no living creature. One day the cubs went for a walk and came upon a ruin. They entered and saw on the wall of the deserted castle a picture of Samson tearing a lion cub in two. Immediately they ran to their father crying out, "We have seen a creature stronger than ourselves and we are frightened." The old lion questioned them and on learning what they had seen, he said, "This picture should assure you the race of lions is the strongest of creatures, for when once a stronger creature appears, it is pictured as a miracle; exceptions prove the rule."

Commenting on Psalms 104:24 the Midrash says, "It happened that a lion, dog, and an Ethiopian gnat were together. The lion was about to mangle the dog, but when he saw the Ethiopian gnat, he drew back in fear for the gnat is the scourge of the lion even as the dog is the scourge of the Ethiopian gnat. Thus the three creatures did no harm to one another. Thereupon Rabbi Akiva quoted 'How manifold are thy works O Lord' " (*Yalkut Shimoni, Midrash Tehillim* 862).

One of the illustrious sons in Israel was Rabbi Isaac Luria (1534–1572). He was a great kabbalist and mystic and was designated as a teacher. Because of his German origin he was known as the ARI (lion), an abbreviation for Ashkenazi Rabbi Isaac. He eventually settled in Safed and became engrossed in the *Zohar*. He lived the life of an ascetic and dissociated himself from the norms of life. Hence he was known as the

Ari HaKadosh, the holy lion. He emphasized to a great degree *kavanah* (devotion to prayer) and performance of *mitzvot*.

Proverbial Sayings

A living dog is better that a dead lion (Ecclesiastes 9:4).

Be a tail to lions and not a head to foxes (*Avot* 4:23).

A handful doesn't satisfy the lion's hunger (*Berachot* 50b).

Lizard

The lizard, of which there are many species, is one of eight reptiles that creep on the earth and defile, as mentioned in Leviticus 11:29–30.

There is some confusion as to which Hebrew word means lizard; some translate *anaka* and others *koach* as "lizard," but it is generally conceded that *haltaah* means lizard. According to Jastrow, the first letter of *haltaah* is not the definite article but is part of the word, the etymology of which is unknown. It should be noted that the word *semamit* in Proverbs 30:28 is translated as spider or lizard.

In the Midrash we have the expression "*ben hanefillin*," literally "son of giants," which typifies a water lizard; it must have been large to fit its name (*Exodus Rabbah* 15:28).

The lizards of Mehuza, which are dry, are unclean if their shapes are retained (*Niddah* 56a), and in another passage we learn that a sand lizard at its first formation is of the size of a lentil when it is whole (*Nazir* 52a).

In *Berachot* 33a there is a story of an *arod*, which is a crossbreed between a snake and a lizard. An interesting story of Chanina ben Dosa and an *arod* is recorded in the Talmud: On learning that an *arod* injured people, Rabbi Chanina placed his foot on the hole from which the *arod* usually emerged, and it promptly bit him, but then it died. He carried it over his shoulder and brought it to the House of Study. He then addressed himself to the disciples, "See, my sons, it is not the *arod* that kills." At the time, this saying originated, "Woe to the man that meets an *arod* but woe to the *arod* that Rabbi Chanina ben Dosa meets" (*Berachot* 33a).

One teacher observed that the convulsive reflex movements in the tail of the lizard continued even after it was cut off (*Bava Batra* 142b).

An interesting incident is recorded in the Talmud where we learn that a motionless lizard was found lying in the Temple abattoir. As a lizard can defile, it was feared that it would contaminate the Temple. Rabbi Gamliel, however, advised that a glass of cold water be thrown over it, and it began to slither along, showing signs of life (*Pesachim* 88b).

The rabbis also inform us that the eggs of the lizard have the white and the yolk mixed together. This is clearly stated in the Talmud: "If the white and yolk are mixed up it is a reptile's egg" (*Avodah Zarah* 40a; *Chullin* 64a), and one rabbi compared the *haltaah* with a weasel because both of them have thick skins (*Shabbat* 107b). According to *Chullin* 127a the salamander is a kind of lizard that was supposed to extinguish fire.

Because the lizard resembles somewhat the serpent, it is disliked but it is useful as it destroys a number of harmful insects.

Commenting on the words "an Egyptian delivered us" (Exodus 2:19), the Midrash recalls that Moses can be compared to one bitten by a lizard, who ran to place his feet in the water. When he put them in the river he observed a small child was drowning so he stretched out his hand and saved him. Thereupon the child said, "Had it not been for you I would have perished." To which the man replied, "Not I have saved you, but the lizard who bit me and from which I escaped." The daughters of Jethro greeted Moses, "Thanks for saving us from the hands of the shepherds." Moses said, "The Egyptian whom I slew; he delivered you." Thereupon they said to their father, "An Egyptian (not Moses), but the Egyptian whom he slew" (*Exodus Rabbah* 1:39).

Proverbial Saying

Immersion in a *mikveh* with a dead reptile in hand (*Taanit* 16a).

Locust

The locust is mentioned often in the Bible and is known by a variety of names.

Arbeh is the most common name by which the insect is known. Egypt was inflicted with the plague of *arbeh*, "locusts" (Exodus 10:4).

The word is derived from the root *ravah*, "to increase, multiply." Some trace it to *arav*, "to attack, devastate." The locust possesses both these characteristics: it multiplies at an alarming rate and is a potential destroyer of crops, which it devours with an insatiable appetite. Commenting on the words in the Torah "And they (locusts) shall cover the face of the earth, that one shall not be able to see the earth" (Exodus 10:5).

After describing how locusts shall cover the earth, the Torah adds these words, "They (locusts) shall eat the residue of that which is escaped, which remains unto you from the hail, and shall eat every tree which grows for you out of the field" (Exodus 10:5). The above statements are no exaggeration for it was reported that in 1978 a full-scale locust plague threatened fifty nations. From Morocco to Pakistan the swarms of locusts laid waste to hundreds of square kilometers of Ethiopia and Somalia and threatened to do the same to Kenya and Tanzania.

When the Midianites attacked the Israelites in the days of Gideon, destroying the produce of the earth and leaving no sustenance, the Bible describes the destruction in this way: "They came in as locusts (*arbeh*) for multitude" (Judges 6:5).

Other words for locusts include the following:

Soleam (Leviticus 11:22), an edible winged insect. Ibn Ezra derives this word from *sela*, "a rock," because it frequents rocks. It is also called the bald locust. The B. D. B. Hebrew Lexicon and Jastrow trace the word to a root meaning "to swallow or ruin," which describes the insect as a swallower and devourer.

Hargol (Leviticus 11:22), according to Jastrow, is a word derived from *chagal*, "to go around," with the letter *resh* inserted. It is translated as "beetle, cricket, or grasshopper."

Hagav (Leviticus 11:22, Numbers 13:33, Ecclesiastes 12:5): according to the B. D. B. Hebrew Lexicon this is probably a nonflying species of locust or grasshopper; its derivation is unknown.

Gazam is derived from a root meaning "to cut off." Thus, Amos exclaims, "The multitude of your gardens and your vineyards and your fig trees and your olive trees has the palmer worm (*gazam*) devoured" (Amos 4:9; see Joel 1:4, 2:25).

Yelek derived from the root *lakak*, "to lick." "The canker worm (*yelek*) spreads itself (licks up the vegetation) and flies away" (Nachum 3:16). "The canker worm in its development sheds the skin that confines its wings and at once flies off" (Joel 1:4, 2:25, Psalm 105:34, Jeremiah 51:14).

Chasil is derived from the root *chasal*, "to finish off or bring to an end." This translation describes well the function of the *chasil*, as we see in the Book of Joel: "That which the palmar worm has left has the locust eaten; that which the locust has left has the canker worm eaten; that which the canker worm has not eaten has the caterpillar (*chasil*) eaten" (Joel 1:4). We readily see that the *chasil* was the most devastating species of locust mentioned in the verse (see also 1 Kings 8:37, Psalm 78:46, Isaiah 33:4).

Tzelatzal is derived from the root *tzalal*, "to quiver." It is translated as a whirring locust (Deuteronomy 28:42). This may be an onomatopoeic designation of locusts in general.

Gav or *gov*: Jastrow traces this word to *gava* and *gavach* and therefore translates the word as "hump-backed *govay*," a species of locust or grasshopper. Compare Nachum 3:17: "Your crowned are as the locusts and your marshals are as the swarms of grasshoppers."

It thus seems that some authorities confuse locusts with grasshoppers because there is a strong resemblance between certain species of both. However, although locusts collect in swarms and migrate, grasshoppers are more or less solitary.

One further biblical passage demands our attention. Solomon declares that the locusts have no king, yet go they forth all of them by bands (Proverbs 30:27). It is possible that Solomon is contrasting locusts with bees, which have a queen that can be distinguished from the swarm because of its size. Yet, locusts have no king that can be differentiated from the rest; they move in companies, are united as one band, and are all alike. The Midrash on the above verse has an interesting reference to Alexander the Great who "in his unrest drove all the world like a locust that flies in the air" (*Yalkut Shimoni*, Proverbs 964).

The locusts are different from all other insects, for although insects are generally forbidden to be eaten, certain locusts are permitted as food as qualified by the Talmud. They must have four feet, two hopping legs, and four wings that are large enough to cover the whole of their bodies (*Chullin* 59a); there is a reference to a clean (kosher) locust in the Mishnah (*Shabbat* 90b).

The Talmud reports that Rabbi Judah ordained a fast when locusts had arrived, but when no damage had been done he exclaimed: Have they then brought provision with them? (*Taanit* 21b).

Locusts are recognized as a valuable source of protein, fat, and calories and contain a fair amount of mineral salts, but they are not rich in vitamins. Locusts are entirely vegetarian in all their stages. This is another reason why they are permitted as food.

In midrashic literature an important item emerges from a comment on the eighth plague. It appears that the true borders between Egypt and Ethiopia were in dispute. It was only after the plague of locusts came that extended to the extremities of Egypt that the correct borders became established (*Exodus Rabbah* 13:4).

Why did God bring the locusts upon the Egyptians? Because they made Israel sowers of wheat and barley, therefore did He bring the locusts that devoured all that the Israelites had sown for them (*Exodus Rabbah* 13:6).

Why did God fix a time for this plague "tomorrow" and not bring it immediately? In order that the Egyptians might feel remorse and do penitence.

Commenting on the words "there remained not one locust in all the borders of Egypt" (Exodus 10:19) Rabbi Johanan declared that at first the Egyptians rejoiced and filled their barrels with them (for food) but even these disappeared (*Exodus Rabbah* 13:7).

In another passage we are informed that the human embryo at the beginning is like a species of locust called *rashon*, which (according to *Chullin* 65b) is another name for *soleam* or *hargol* (*Leviticus Rabbah* 14:8).

Yet another name for a species of locust with long heads is *shoshiva* (*Avodah Zarah* 37a).

God has provided every creature with a self-defense mechanism that wards off a potential enemy. This is especially true of the locusts. It is incredible how they are protected from their main enemy, the birds, who cannot even get near them. At Princeton University in 1970 a body of scientists discovered that the noise generated by a swarm of thousands of locusts is so deafening even at a distance of sixty feet that it has been compared to a pneumatic drill breaking up the stones of a roadway. It is this deafening noise that frightens birds and people from approaching them.

Although this scientific information is fascinating, it is of course anticipated by the Bible. This is how the Prophet Joel described the noise of the locusts: "Like the noise of chariots on the tops of the mountains do they leap, like the noise of a flame of fire that devours the stubble, as a mighty people set in battle array" (Joel 2:5). Many commentators treat the reference to the locusts in Joel in a figurative sense, a flight of the imagination, but acting on the findings of modern scientific thought we have no hesitation in treating this passage literally.

One further problem needs clarification. If the noise locusts generate is so deafening that it can burst an eardrum, how do they manage to live with each other? Nature again solves the problem. Science informs us that when they are together their own eardrums collapse temporarily so that they suffer no injury.

Proverbial Saying

For the crime of robbery, locusts make their invasion (*Shabbat* 32b).

Monkey or Ape

The word *kof*, "monkey," is probably derived from the root *nakaf*, "to go around." Monkeys usually jump and skip around with great alertness. Others suggest that the word *kof* is not Hebrew, but is traced to the Sanskrit *kapi*.

Only the plural form is found in 1 Kings 10:22 and 2 Chronicles 9:21: "The navy of Tarshish brought gold, silver, ivory, apes and peacocks" (for King Solomon). Commenting on the above verse Kimchi remarks that the face of the monkey resembles that of man. This is obviously based on the Midrash that states that the faces of human beings started becoming those of monkeys from the generation of Enosh (*Genesis Rabbah* 23:9).

In the seven stages in the life of man the rabbis compare an old man to one bent like a monkey; that is, he loses the appearance of a human being (*Ecclesiastes Rabbah* 1:3).

This idea of finding a resemblance between an ape and a human being is elaborated on in the Talmud: "Compared with Sarah, all other people are like a monkey to a human being, and compared with Eve, Sarah was like a monkey to a human being, and compared with Adam, Eve was like a monkey to a human being, and compared with the *Shechinah*, Adam was like a monkey to a human being" (*Bava Batra* 58a).

Ishmael said, "It is allowed to breed apes because they keep the house clean" (*Bava Kamma* 80a).

The Mishnah has an interesting reference to wild manlike creatures, which it is conjectured may be chimpanzees or gorillas and is a translation of *avnei hasadeh* in Job 5:23 (*Kilayim* 8:5).

The Talmud narrates the following story in the name of Rabbi Gamada. He gave four zuzim to some sailors to bring him something. But as they could not obtain it they brought him a monkey for it. The monkey escaped and made his way into a hole. In searching for it they found it resting on precious stones and brought them all to him (*Nedarim* 50b).

A scientific experiment performed by Harlow in the 1950s is worthy of mention. Two orphaned monkeys were placed in a laboratory and fed by bottle. Periodically they were cradled in the arms of an electrically warmed stuffed monkey. After a while they became closely attached to the stuffed "monkey." They had but one failing–they lacked the ability to reproduce. Therefore, artificial warmth with technology may keep the patient alive only in a robotic form of existence.

It is reported that Robert M. Yeaks, professor of psychology at Yale University, in 1943 said: "If in captivity plenty of room is given to the chimpanzee, sex relations are restricted to a few days during their monthly cycle. This seems to follow the same pattern as in Jewish Law."

Because of the close resemblance between man and monkey the evolutionists have consistently used the monkey to prove their thesis that man has evolved from the monkey and was not created by God! The monkeys of Japan are ballyhooed as evidence of evolution. "These carefully studied creatures appear to be following a historical path breathtakingly like our own." Breathtaking indeed! The writer admits that "there are, of course, tremendous differences between these animals and man." But he is confident that matters will straighten themselves out. "Yet in their social behavior Koshima's whip-smart monkeys do seem to be closing the gap." Soon it seems they will become who knows what. But the question presents itself: Why have these monkeys waited so long? What were these monkeys doing until now?

The sad truth, however, is that if we could observe them 5,000 years later we would see that they are still nothing but monkeys just as were their ancestors 5,000 years ago when they were first created.[1]

Proverbial Sayings

"O what ugly creatures there are in the world," sighed the monkey (J. Cahan, Tel Aviv, 5710).

Men's faces were made to be apelike (*Genesis Rabbah* 23).

As inferior in looks as the ape to man (*Bava Batra* 58).

Mosquito (Gnat)

The Hebrew *yatush* is translated by Jastrow as "mosquito or gnat." It is not found in the Bible, but it should be noted that the fourth plague with which the Egyptians were smitten, *arov*, is translated in the Midrash as "a swarm of wasps and gnats" (*Exodus Rabbah* 11:4), whereas the B. D. B. Hebrew Lexicon renders it as "stinging flies."

About one hundred years ago a British doctor, Sir Ronald Ross, first recognized under the microscope the parasite of malaria in the tissues of a dissected mosquito. As malaria is prevalent in hot eastern countries such as Egypt, it is perhaps reasonable to suggest that the *arov* and *yatush* are a species of mosquito and gnat.

The rabbis often emphasized the virtue of humility, and they warned against boastfulness. Thus at the creation, man, who is superior to the animal, was reminded that the insect came before him.

If a man does worthily, they say to him: "Thou wast created before the angels of the service." If he does not, they say to him: "The fly, the gnat (*yatush*), the worm were created before you" (*Genesis Rabbah* 8:1).

The Talmud expresses the same idea this way. Why was man created on Friday? So that if he became overbearing, one can say to him, the gnat (*yatush*) was created before him (*Sanhedrin* 38a).

The Talmud further states that the Holy One did not create anything without purpose. Thus He created the mosquito (crushed) for an antidote to a serpent's bite (*Shabbat* 27b).

Moreover, we learn from our rabbis that there are five instances of fear cast by the weak over the strong. One of them is the fear of the small mosquito by the large and powerful elephant. This is caused by the mosquito entering the trunk and overpowering the elephant (*Shabbat* 27b).

In *Sanhedrin* 77a, it is reported that the deadly mosquito can sting a person to death. Rabbah drove away mosquitoes not to harm living creatures.

The story of the wicked Titus who blasphemed the God of Israel and challenged Him to fight on land is recounted in the pages of the Talmud. When Titus came on dry land, a voice came forth from heaven saying: "I have a tiny creature in my world called a gnat. Why is it called a tiny creature? Because it has an orifice for taking in but not for excreting. Go up on the land and make war with it." When he landed, the gnat came and entered his nose and it knocked against his brain for seven years. . . . When he died they split open his skull and found something like a sparrow two selas in weight" (*Gittin* 56b).

"The mosquito has more efficient signaling equipment than any radio system of communication. Merely by vibrating its wings it can set up a humming that will penetrate thunder, sirens or any other jamming noises and deliver its message two hundred feet away."[1]

A story is told of David, king of Israel, who wondered as he lay on his couch hearing a mosquito buzz, "Of what use is the mosquito? What can he be good for? He only disturbs our comfort." The rabbis say that King David was taught that there was nothing in the world without its use. David once entered the camp of Saul to take away the king's weapons. Abner, the servant of the king, was sleeping nearby, and as David stooped, Abner moved in his sleep and placed his leg upon David's body. Now if David had moved, Abner would have awakened and killed him for entering Saul's tent; and if David waited until morning, certainly death awaited him. Suddenly a mosquito alighted on Captain Abner's leg, and he moved it. David quickly escaped and realized that a mosquito had its use.

Proverbial Saying

The mosquito (an application of a pulp made of mosquitoes) is remedy for a serpent bite. (*Shabbat* 77b).

Moth

The moth and the butterfly are two divisions of the same family of insects, of which there are hundreds of varieties in Israel. The moth moves around at night, whereas the butterfly with its beautifully colored wings operates in the day and uses the wings with which the Almighty endowed it. The Hebrew word for "moth," *ash*, is derived from a root *ashash*, which means waste away or disintegrate (compare Psalms 6:8, 31:10).

The moth is small but very destructive, especially of clothes and woolens. Thus Isaiah exclaims: "Behold they all shall wax old as a garment, the moth shall eat them up" (Isaiah 50:9). *Sas* is synonymous with *ash* and occurs in Isaiah 51:8 where both words are found in the same verse: "for the moth shall eat them up like a garment and the worm shall eat them like wool."

In Isaiah 10:18 the difficult expression *kimsos noses* is normally translated as "when a sick man wastes away." Others connect *noses* with *sas* meaning "a moth" and so render the verse as follows: "The trees of the forest shall be destroyed as if from decay caused by the moth."[1]

The moth symbolizes decay and destruction. "There am I unto Ephraim as a moth and to the House of Judah as rottenness" (Hosea 5:12). "As the moth and rottenness do their work in silence, so will God's retribution stealthily come upon them" (Soncino Bible). The moth invades not only garments as in Job 13:28, "like a garment that is moth-eaten," but it also attacks houses of clay: "them that dwell in houses of clay whose foundation is in the dust, who are crushed before the moth" (Job 4:19). The moth will gnaw even at the foundation of the home and bring it down. This is reminiscent of the wicked who builds his house as a moth (Job 27:18).

In the Talmud, reference is made to scrolls of the law and mantles that are moth-eaten, and the rabbis advise that they should not be cast aside but "hidden," that is, stored away in a *genizah* and eventually buried (*Shabbat* 90a).

Commenting on the word *erez* (cedar) in Song of Songs 8:9 the rabbis say it is *sasmagor*, a combination of *sas* and *magor* that according to Jastrow is a sawing worm, a name of a species of cedar subject to decay (*Yoma* 9b).

The Midrash has a striking simile regarding Moses' inability to enter the Land of Israel. "All the delight which Moses longed for, to enter the land—you have caused it to decay as a moth enters garments and makes them decay" (*Deuteronomy Rabbah* 2:2).

Rabbi Bunam was asked by the Lubliner Rabbi for a remedy to stop the hair from falling out of his fur hat (*streimel*). Rabbi Bunam answered: "If all the moth-eaten hair is combed away the remainder will not fall out, otherwise the good hair will also fall prey to the moths." "Let this be a lesson to you, Bunam," explained the Lubliner, "when you become a leader in Israel" (*Ramatha'im Tzofim*, Shinaver).

Mouse

The mouse, *achbar*, is mentioned several times in the Bible and is a generic name for small rodents, of which there are twenty-two species in Israel. The origin of the name *achbar* is doubtful, but it is noteworthy that in Greek, Latin, and English the name is almost the same, and this suggests that it was very common.

The mouse is first mentioned in Leviticus 11:29 where it is included among the unclean creeping things (*sheratzim*). As a proper name, *Achbar* is found in Genesis 36:38 and 2 Kings 22:12.

Mice are prolific and reproduce several times a year, from five to ten at a litter. Herodotus informs us that the army of Sennacheriv, king of Assyria, was once defeated because of mice that in one night gnawed at the harnesses to such an extent that they were compelled to retreat.

Commenting on the story of the mice in 1 Samuel 6:4, Maimonides says the chief object was the removal of the injury caused by the multiplicity of mice that ravaged and destroyed the crops in the fields of the Philistine cities.

In Isaiah 66:17 we read of those who "eating swine's flesh, and the detestable thing, and the mouse, shall be consumed together." Some think that "the detestable thing" here is the weasel, which is mentioned together with the mouse in Leviticus 11:29.

The Midrash on the above verse in Isaiah asks why is the prohibition of swine stricter than that of other unclean beasts or that of the mouse stricter than that of other creeping things? Truly, it is not. Both the swine and mouse were mentioned, but the same applies to all other unclean cattle and creeping things.

In rabbinic literature we learn some interesting facts about the mouse. For instance, it is not promiscuous: God does not withhold the due reward of any of his creatures, even of the mouse, which preserved its family and mated with its kind, not like the men of the generation of the flood who were promiscuous (*Tanchuma*, *Noah* 18b, Buber edition).

The story is told of Rabbi Pinchas ben Jair who went to a place where they complained that the mice had eaten their grain. He summoned the mice and they squeaked. He asked the men if they understood what the mice were saying and promptly explained that the mice were aggrieved that the grain was not properly tithed. Whereupon they said to him, "Pledge yourself to us that the mice will not eat our grain if we tithe it properly." He pledged himself to them, and they suffered no longer (*Demai* 22a).

In the Talmud the rabbis analyze a case in which a mouse fell into a cask of beer and Rav prohibited the beer from being drunk because the mouse is something repugnant and people recoil from it. The rabbis also discuss the case of a mouse falling into vinegar, which is sharp and pungent, and so the mouse would not affect its taste. Here Rabbi Ashi prohibited the vinegar from being eaten because the mouse might have been dissolved and there was the fear that one may swallow a piece of the mouse, which is forbidden (*Avodah Zarah* 68b–69a).

The rabbis declared that the mouse is part flesh and part dust and in its development becomes all flesh (*Sanhedrin* 91a). This belief that mice can develop from the earth is upheld by Maimonides who, in his comment on *Mishnah Chullin* 9:6, states that many people have claimed to have seen a mouse made of part earth and part clay.

Several characteristics of the mouse are detailed in the Talmud. We must distinguish between a sea mouse (which is a fish resembling a mouse) and a land mouse: the former conveys no uncleanness, whereas the latter does (*Niddah* 43b, *Chullin* 126b). Mice do not seize food from each other (*Pesachim* 10b). A mousetrap is not subject to uncleanness

(*Kelim* 15:6). The mouse causes destruction to clothes and wood without any profit to itself (*Horayot* 13a). The mouse will attack even a human corpse (*Shabbat* 151b).

Terumah, the sacred food, used to be placed near the Scroll of the Law, as both are holy. When, however, the food attracted mice the rabbis imposed uncleanness upon the sacred books (*Shabbat* 14a). In the case of a mouse, a proportion of one to one thousand was required to neutralize it (*Yerushalmi*, *Terumah* 47).

The Kobriner Rabbi commented on Psalm 10:10: "He crouches, he bows down and the helpless fall into his mighty claws." In explanation he told the following fable: An old mouse sent out her son to search for food, but warned him to be careful of the enemy. The mouse met a rooster and hastened back in terror. He described the enemy as a haughty being with an upstanding red comb. "He is no enemy of ours," cried the mouse and sent her son out again. This time he met a turkey and was even more terrified. "Neither is he our enemy," said the mother, "our enemy keeps his head down like an exceedingly humble person: He is smooth and soft spoken, friendly in appearance and acting as if he were a very kindly creature. If you meet him beware!" (*Or Yesharim*).[1]

Proverbial Sayings

A man was nicknamed "The mouse lying on the denarii" (that is, a miser) (*Sanhedrin* 29b).

It is not the mouse that is the thief, it is the hole . . . if there were no mouse how should the hole come by it? (*Gittin* 45a).

Mule

The mule is the result of the union of a stallion and a she-ass. Though the Torah forbids crossbreeding, the mule is permitted to be used as a beast of burden.

Indeed, the mule was very popular with princes and the aristocracy. In 2 Samuel 13:29 we read: "Then all the King's sons arose and every man got him up upon his mule and fled" (compare 2 Samuel 18:9). In 1 Kings 1:38 we learn that Solomon rode upon King David's mule, and the prophet Ezekiel informs us that they of the House of Togarmah traded with horses, horsemen, and mules (Ezekiel 27:14).

The general Hebrew word for "mule" is *pered*, which some connect with a Syrian root meaning to flee. Especially in mountainous regions and steep declines, the mule does not stumble but easily slides along.

Pered is found often in the Bible and in Psalm 32:9 we are reminded that man is morally superior to the horse and mule, who can be contained only by bit and bridle.

The Hebrew word *yemim* (Genesis 36:24), which is often translated as "hot springs" connected with *yam*, "sea," was rendered by the Talmud as "mules" that terrify human beings (*Chullin* 7b). *Targum Jonathan* to Genesis 36:24, *Berachot* 8:5, and *Genesis Rabbah* 82:15 all render *yemim* as "mules," where they are referred to as *hemionos*, the Greek form of mule.

In the Talmud we read that Hormin (the name of the demon Ormuzd) saddled two mules that stood on the two bridges of the river

Rognag. He jumped from one to the other backward and forward, holding in his hands two cups of wine and pouring alternately from one to the other. Not a drop fell to the ground (*Bava Batra* 73a, 73b).

The rabbis homiletically interpret the word *adrammelech* (2 Kings 17:31) to mean that the mule honors its master or king (*melech*) by carrying his burden (*Sanhedrin* 63b).

Though the mule is a hybrid, one teacher, Rabbi Nechemiah, is of the opinion that the first mule was created directly at twilight on the eve of the Sabbath at the creation of the world (*Pesachim* 54a). R. Simeon ben Gamliel said: "The mule came into existence in the days *Anah*" (Genesis 36:24). Those who interpret *Anah* symbolically used to say: *Anah* was unfit (the issue of an incestuous union). Therefore he brought unfit animals into the world; in other words evil begets evil (Genesis 36:24).

The *Halachah* ruled that if a mule was craving for sexual gratification it must not be mated with a horse or an ass (contrary to the laws of the Torah), but only with one of its own species (*Ketubot* 111b).

Ostrich

The ostrich is known by three different names: *renanim*, "a bird of piercing cries"; *yaen*, "a bird of greed"; and *bat hayaanah*, "a daughter of voracious greed." Feliks, however, is of the opinion that *bat hayaanah* is a mistaken identification of *yaen* and does not mean ostrich.[1]

The ostrich is mentioned in Lamentations 4:3 and in Job 39:13–18 where a full description of its habits is given. Here it is called *kenaph renanim*. When Job says, "Its wings beat joyously," he refers to the large feathers on the wing and tail that alternate in white and black and are used for decorative purposes. These feathers are unique in that they are not joined to each other but are strong, independent, and separate. Nor are the feathers used for flying, but to help ostriches run swiftly. Alternatively, they are used as fans, as ostriches generate much heat.

From the second half of verse 13, chapter 39, and further Job seems to contrast the affection of the kindly stork for her young with the cruel tendencies of the ostrich, which in its greed leaves her eggs to be trampled upon: "She is hardened against her young ones as if they were not hers."

An alternative view is given by one writer who declares, "But most naturalists confirm the statement of the natives that the eggs on the surface are left in order to afford sustenance to the newly hatched chicks which could not otherwise find food at first in these arid regions."[2]

In Job 39:17 Job speaks of the ostrich as being deprived of wisdom and understanding, and in verse 18 we learn that such is the speed of this bird that it can outdistance a horse and its rider.

In its voracious appetite the ostrich eats almost everything, and according to the Midrash it was fed on glass during its stay in the ark (*Genesis Rabbah* 31:14, *Shabbat* 128a).

The Mishnah informs us that the eggs of the ostrich were turned into vessels (*Kelim* 17:14). This is not surprising because we know that the shells are very hard, and the eggs are sometimes more than five inches in diameter and weigh more than twelve pounds.

In Mishnah and Talmud the name *naamit* is used for "ostrich."

Owl

The owl is a bird of prey that sleeps during the day and prowls at night. Owls do not have nests and are usually found in ruins and desolate places.

The Bible mentions by name each of the different species of owl, and they are characterized by their freakish features, which have given them demonic powers. As these species are members of one family, we shall begin with the various names listed in the Torah.

Bat hayaanah is found in Leviticus 11:16 and in Deuteronomy 14:15. *Bat hayaanah*, the name of this owl, has been translated by Fisher as "the daughter of one that answers," because in their hootings the owls answer one another.[1] Feliks asserts that the name ostrich, often given to this bird, cannot be accepted, as it is mistaken for *yaen*, the recognized name for ostrich.[2] The *bat hayaanah* lives among the ruins, whereas the ostrich lives in the open places. The Talmud argues that *yaanah* and *bat yaanah* are used indiscriminately (*Chullin* 64b).

The Septuagint renders the owl as *ulula*, which in English means to "ululate, to howl, or lament." Note the words of the prophet Micah: "I will make a wailing like the *tanim* and a mourning like the *bnoth yaanah*," which Feliks[3] translates as "dark desert eagle owl" (Micah 1:8).

The next member of the owl family found in the Torah is *kos* (Leviticus 11:17, Deuteronomy 14:16). This word is generally rendered as "pelican or little owl" as it is the smallest of this species. In Deuteronomy 14:17 it is written as *kaat*.

This bird is found among ruins and tombstones and groans in a melancholy tune. Thus the psalmist says: "I am like a pelican (*kaat*) of the wilderness; I am become as an owl (*kos*) of the waste places" (Psalm 102:7). One writer called it the most sombre, austere bird he ever saw.

The owl is called by the Arabs "mother of ruins" because it dwells among ruins and lonely places.

The rabbis declared that to dream of any kind of bird is a good omen, with the exception of the owl and horned owl because of their association with derelict regions (*Berachot* 57b). The rabbis also refer to two species of owl by the name of *kadia* and *kinufa*, which are characterized by eyes that are placed in front of the head and that have jaws like those of human beings (*Niddah* 23a).

The *kos* is followed by the *shalach*, which belongs to the tribe of owls, but as it alone feeds on fish, Feliks calls it the fish owl.[4] English versions of the Bible call it the cormorant. However, one writer suggests that the cormorant dives only from the surface, whereas other birds dive and plunge from a height, and the *shalach* is therefore a generic term for plunging birds.[5]

This title, "plunging bird," would explain the derivation of *shalach* from a root meaning "to cast or hurl." It hurls itself into the sea and catches the fish with its legs. "The soles of its feet have a rough surface to prevent the fish from slipping out."[6]

Rashi on Leviticus 11:17 quotes the Talmud, which derives *shalach* from a root meaning to draw or pull out: "That is the bird that catches fish out of the sea" (*Chullin* 63a).

The word *yanshuf* (Leviticus 11:17 and Deuteronomy 14:16) is derived from *neshef*, "darkness or twilight," the time during which this nocturnal bird looked for and hunted its prey. The B. D. B. Hebrew Lexicon calls it a bird with hard strident notes. Isaiah confirms that the *yanshuf* dwells in the wasteland (Isaiah 34:11).

This species of owl has been called the great owl, the Egyptian owl, and bittern, and Feliks calls it the long-eared owl. Rashi on Leviticus

11:17 describes it as a bird that shrieks at night and has a jaw resembling that of a man. Ibn Ezra remarks that the *yanshuf* flies in the dark because the light of the sun or the light of day dazzles its eyes.

Tinshemet (Leviticus 11:18; Deuteronomy 14:16) is derived from the root *nasham*, "to pant or breathe." The B. D. B. Hebrew Lexicon says it is reminiscent of the deep and strong breathing of a woman in travail. It is considered to have demonic powers because of its frightful voice and unusual features.

The *tinshemet* is mentioned twice, once here, among the birds, and again among the creeping things in Leviticus 11:30.

Rashi, in commenting on Leviticus 11:18, sees a resemblance between the *tinshemet* and the mouse. The Talmud calls it *bavat*: the *bavat* among the birds and the *bavat* among the creeping things (*Chullin* 63a). The rabbis also observe that the most repellent among the birds is the *tinshemet*, and the most repellent among the creeping things is the *tinshemet*.

The English versions of the Bible render it as "horned owl, swan, and mole." However, the *tinshemet* is unlikely to be the swan, as the duck and goose, which belong to the swan family, may be eaten, whereas the *tinshemet* is listed among the prohibited birds.

In addition to the species described above, which are all found in the Pentateuch, several species are found in the other books of the Bible. *Oach* is found only in Isaiah 13:21. The English versions of the Bible render it as "doleful creatures," but the Jewish Publication Society translates it as "ferrets." This word has also been translated as "jackals, martens, and monkeys." The name is onomatopoeic, reflecting the mournful sound it gives forth. It is a large predatory bird, and it dwells among the ruins, as do so many of the owl family. Felix calls it the eagle owl.

Lilith (Isaiah 34:14) is translated by the English versions of the Bible as "screech owl, night monster, and lilith." Feliks calls it the tawny owl. It is a night bird, as the Hebrew *lilith* refers to the night. In the Talmud *lilith* is introduced as a female demon of the night reputed to have wings and a human face (*Niddah* 24b, *Bava Batra* 73a, *Eruvin* 100b).

English versions of the Bible translate *kipod* as "porcupine or bittern" and *kipoz* as "arrowsnake"; Feliks calls it the short-eared owl. The B. D. B. Hebrew Lexicon traces the Hebrew *kipod* to the root meaning "to roll up." Both the porcupine and this species of the owl curl up when resting on the ground.

The *kipod* is found three times in *Tanach*: in Isaiah 14:23, 34:11; in Zephamia 2:14. It has been suggested that ruins and desolate places, which these birds inhabit, were supposed to be occupied by spirits and demons; the word *kipod* therefore may refer to a demon. In modern Hebrew a *kipod* is a hedgehog. This follows the Septuagint and Vulgate translations.

The *kipoz* occurs in Isaiah 34:15 as a token of desolation. The B. D. B. Hebrew Lexicon and Jastrow compare *kipoz* with *kafatz*, "to spring, jump," and quote one authority who suggests that the habit of the arrowsnake is to leap from trees onto passersby.

The spraying of pesticides to rid the crops of dangerous pests is universally practiced. However, recently it has been discovered that some elements of the pesticide may attack even the heart of the fruit, which poses a very dangerous health risk. This is a typical example of the ineffectiveness of the artificial aids introduced by man as compared to nature, which has the correct answer. It is well known that the barn owl is the best and most efficient destroyer of all rodents that invade the crops of the field.

Ox and Cow

In ancient Palestine the domestic ox was a very important animal and had various uses, including plowing as a beast of burden, and as a sacrifice.

"Thou shalt not plow with an ox and an ass together" (Deuteronomy 22:10). The law applies to all animals, but the ox is mentioned because it was essential in biblical times. Indeed, it was part of the human community, and it earned its weekly day of rest: ". . . . on the seventh day thou shalt rest: that thine ox and thine ass may rest . . ." (Exodus 23:12).

Oxen were used as beasts of burden, as reflected in these biblical passages:

"Brought bread on asses, and on camels, and on mules, and on oxen" (1 Chronicles 12:41).

"So the Lord blessed the latter end of Job more than his beginning: for he had . . . a thousand yoke of oxen" (Job 42:12).

"Joshua says to the tribes of Reuben and Gad: 'Return with much wealth unto your tents, and with very much cattle, and with silver and gold.' Cattle is placed before silver and gold" (Joshua 22:8).

Because natural fodder was not available throughout the year it was difficult to maintain cattle: "And on all hills that shall be digged with the mattock, there shall not come thither the fear of briers and thorns: but it shall be for the sending forth of oxen, and for the treading of sheep" (Isaiah 7:25). Cattle were fed in stalls: "Better is a dinner of herbs where love is, than a stalled ox and hatred therewith" (Proverbs 15:17).

Oxen were offered on the altar as sacrifices (Leviticus 17:3–4; Numbers 7:17, 21). The ox was a sacred animal, particularly to the Babylonians and Egyptians, and the Israelites had to be weaned away from such worship. In the Temple of Solomon the large laver of wash-basin was supported by figures of twelve oxen (1 Kings 7:25, 29, 44).

The Bible details several characteristics of the ox. Though the ox is generally gentle and patient, some have a tendency to gore and cause death: "If an ox gore a man or woman" (Exodus 21:28; compare Psalm 22:13).

At a certain age the ox, a clean animal, was slaughtered for food (Leviticus 17:3, Deuteronomy 14:4, 1 Samuel 14:34). Moses compares Joseph to a strong ox (Deuteronomy 33:17). Isaiah likens the instinctive intelligence of the ox to the ignorance of man: "The ox knoweth his owner . . . but Israel doth not know, my people doth not consider" (Isaiah 1:3).

The Bible uses different terms for cattle according to their age, sex, and climatic conditions. One such term is *eleph* or *aluph* (Deuteronomy 7:13, Psalm 144:14). Some connect this word with *aluph*, "a leader," as the ox leads the cattle.

Abir, meaning "strong, mighty," can refer to animals; compare *abirei Bashan*, the strong bulls of Bashan (Psalm 22:13; 68:31).

The most often-used term for cattle and herd ox is *bakar*. Some connect *bakar* with the root "to seek," suggesting that the ox was sought after by man for its many uses in the field and beyond. The ox was not only employed for drawing the plow and wagon but its flesh was widely consumed, its hide tanned into leather for shoes, its horns made into combs and knife handles, its bones used for buttons, its bones and dung, when powdered, for manure; its hair mixed with mortar used for plastering, and its suet made into candles. Discussing the uses of oxen Rabbi Hisda advised that the most valuable hides come from black oxen and the choicest meat from red oxen, whereas white oxen are most fit for plowing (*Nazir* 31b).

From the Talmud we learn that when working with oxen in the vineyards for plowing, and during the harvesting of grapes, much damage may be done to the vines and the oxen may be injured if great care is not taken (*Bava Metzia* 30a).

In Babylonia whenever a question about plowing was concerned, the ox is always mentioned, whereas concerning threshing, the cow is given as an example. Rabbi Joseph declares that the ox may eat of the corn he is threshing (*Gittin* 62a).

Several rabbis specialized in the breeding of animals. Thus we learn of the calves belonging to Rabbi Huna and that Rabbi Hamnuna possessed oxen, some of which strayed from his home (*Sanhedrin* 61a).

A fine ethical concept emerges from a rabbinic interpretation of a scriptural verse. Commenting on Exodus 21:37, "If a man steal an ox or a sheep and kill it, or sell it, he shall pay five oxen for an ox and four sheep for a sheep," Rabbi Meir said: "See how beloved is labor to Him who spoke and the world existed. Fivefold compensation is exacted for the ox because the thief interrupted its labor. Since the lamb does not labor, the compensation is but fourfold." Rabbi Johanan ben Zakkai said: "See how God cares for human dignity. The ox walks on its own feet, but the lamb is picked up and carried" (*Bava Kamma* 79b).

The ox was honored in the beautiful and moving procession of first fruits as they were carried shoulder high to the Temple in Jerusalem. At the head of the procession was the ox with its horns overlaid with gold and a wreath of olive leaves on its head (*Mishnah Bikkurim* 3:3).

The most extensive discussions in the Talmud regarding the ox are formulated in *Bava Kamma*, the first tractate in the order of *Nezikin*, "Damages." The first mishnah of this tractate states that there are four principal categories of damage, and the first is designated as "ox." Thus, the Torah declares that if an ox killed a person it shall be stoned (Exodus 21:28).

The Torah distinguishes between an animal that was "wont to gore" and one that is patient and gentle and does not gore. The latter was known as *tam*, "simple, innocent," whereas the former was termed *muad*, "the ox of an owner who was forewarned." The Talmud defines the *muad* as an ox that has done damage three times and for whom three warnings have been given to the owner. In contrast, the *tam* is an animal that does not gore when it is patted by a child. This important distinction became the norm in Jewish law and was eventually applied to all cases of tort.

The Midrash narrates the following story: An ox was once being led to the altar, but would not let himself be drawn along. A poor man came with a bundle of endives in his hand. As he held them out to the ox, it ate them, sneezed, and expelled a needle through its throat. If the needle had remained it would have caused a perforation, thus invalidating it as a sacrifice. The ox in satisfying its hunger allowed itself to be drawn to the altar (*Leviticus Rabbah* 3:5).

In the Midrash on Psalm 101 Moses asked a hypothetical question regarding the offering of the princes in Numbers, chapter 7: " 'Can some prophet have decreed that the princes bring these offerings or did the princes bring them of their own free will?' The Holy One replied: 'Take it of them in order that the wagons may serve for carrying the tent of meeting' " (Numbers 7:5).

How long did the oxen remain alive? Until the children of Israel came to Gilgal (Hosea 12:12). One teacher said: "Until the Temple was built and Solomon offered them as a sacrifice" (2 Chronicles 7:5). Note it is not written "a sacrifice of oxen," but "a sacrifice of the oxen" that the princes brought for carrying the Tabernacle.

The biblical expression "bring here" is used of a ram, but not of an ox in order that a man should not say he will go and do things that are unseemly and then offer an ox with much flesh, imagining that he has found favor in the eyes of God who will accept him as a penitent. The sacrifice of a ram, though much smaller, is no less efficacious in bringing man near to his Maker (*Leviticus Rabbah* 2:11).

In an interesting comment on 1 Samuel 6:12, "And the cows took the straight road," the rabbis play on the word *yisharnah* and connect it with the word *shir*, "song." They thus interpret the verse to convey this exposition – The cows turned toward the ark and offered up praises to God (*Genesis Rabbah* 54:6).

The Lizensker Rabbi spoke on the text (Exodus 21:33): "And if a man shall open a pit, or dig a pit, and not cover it, and an ox or ass fall therein, the owner of the pit shall make it good, he shall give money to the owner of them." If the *tzaddik* makes an opening in the heart of his followers or if he strives to dig an opening among the hard-hearted and continues his efforts lest they grow harsh again, thereby overthrowing the spirit of worldliness, then the Lord, who has endowed the *tzaddik* with the power to influence human hearts, shall grant him his reward. "He shall repay him by vouchsafing unto him the longing to increase his loyal devotion to the Lord." In this context, the Hebrew root *kasaf* means both "money" and "longing."

The Sassover Rabbi said: Abraham was not commanded to offer up a ram instead of his son Isaac, but he was a prophet, and he foresaw that his children would worship a (golden) calf and he offered up a ram as an atonement for them in advance. An allusion is found in the verse, Genesis 22:13: "and behold, behind him a ram caught in a thicket by his horns." The Hebrew word for thicket is composed of the letters *samech*, *bet*, and *kaph*; the letters following each of these respectively, in the alphabet, are, *ayin*, *gimmel*, and *lamed*, which make up the Hebrew word *egel*, "calf."

Parah, "cow," is the feminine form of *par*, "bullock." This word occurs in the Bible especially in the story of the dream of Pharaoh, (Genesis 41:1) and the law of the red heifer (Numbers 19:2); it also appears figuratively in Hosea 4:16: "for Israel is stubborn like a stubborn heifer."

Today the cow is exclusively known for the milk, cheese, and butter it provides, but in the early days the cow was also used for threshing. Thus, the question is raised in the Talmud whether a Jew may tell a heathen to muzzle a cow and then thresh grain with it. Rabbi Sheshet asked if one may muzzle a cow while threshing to prevent it from eating a kind of grain that will have a harmful effect upon it (*Bava Metzia* 90a). It appears that the cow was also used for plowing in Palestine.

In Midrashic literature the cow is often employed in parables. Discoursing on the suffering of the righteous, Rabbi Eleazar said: "Like a man who had two cows–one was strong and the other was weak. Upon which does he put a burden? Upon the strong. So God does not try the wicked, for they could not endure it, but he tries the righteous" (*Song of Songs Rabbah* 2:1).

Commenting on Genesis 39:1, "And Joseph was brought down to Egypt," the Midrash declares he may be compared to a cow that was resisting being dragged to the abbatoir. What did they do? They drew her young one before her whereupon she followed unwillingly. In the same manner Jacob should have gone down to Egypt in chains, but God said that as he was His first-born son He could not bring him down in disgrace (*Genesis Rabbah* 82:2).

The young of the cow is the calf, *egel*, which is popularly connected with the golden calf. The Talmud does mention that Rabbi Hanina and Rabbi Oshiah spent every Sabbath eve studying the *Sefer Yetzirah*, the book of creation (a mystical book on creation), and by means of this treatise they created a third grown calf and they ate it.

The expression "third year" has been interpreted to mean that the calf reached one-third of its full growth in its third year–third born, fat (*Sanhedrin* 65b).

Proverbial Sayings

When an ox falls, many sharpen their knives (*Shabbat* 32a, *Lamentations Rabbah* 5.1:34).

You cannot get two skins off one ox (Yiddish proverb).

More than the calf wishes to suck, does the cow desire to suckle (*Pesachim* 112a).

According to the size of the ox is the feast. As you call Esau (Rome) great, so will his punishment be (*Genesis Rabbah* 65).

A miserly man and a fat cow are useful only after death (Sholem Aleichem).

Rather one cow in the stall than ten in the field (J. Bernstein, *Yiddish Proverbs*).

While the rope is in your grip, pull your cow (Arama, *Akedat Yitzchak*).

Partridge

The partridge is a small bird known by the name *kore*, "one that calls." This may allude to the call of the male to his mate.

One writer remarks that when the mother is suddenly surprised, she utters a peculiar note at which all her young, a dozen or more, immediately hide and cannot be found. The mother then flutters round, with her feathers drawing her pursuer farther away from the place of her young. When the mother is at a good distance and the danger is no longer apparent, she returns to the place of her young and with her peculiar call she assembles them together in safety.[1]

The partridge has always been a game bird hunted by many. Thus, David complains that King Saul is hunting him, "For the King of Israel is come out to seek a single flea as when one does hunt a partridge in the mountains" (1 Samuel 26:20).

Several partridges sometimes lay their eggs in one nest. As the female is a small bird and does not cover them all as she sits on them, some of the eggs will not be hatched. The Prophet Jeremiah therefore uses the partridge to illustrate the unethical acquisition of wealth, which the rich may not enjoy: "As the partridge that broods over young which she has not brought forth, so is he that gets riches and not by right; in the midst of his days he shall leave them and at his end he shall be a fool" (Jeremiah 17:11).

In Judges 16:19 the place name *ein-hakore* is translated by some as "the partridge spring." This fits in with the notion that partridges are commonly found near springs and swampy ground.

The partridge is a ritually clean bird. Indeed its flesh is very tasty, and the Mishnah records that the male also hatches the eggs (*Chullin* 138b, 140b; *Tosaphot* on 63a, beginning with the word *netz*).

Feliks mentions another species known as *choglah*, which is called "chukar partridge." This is not mentioned specifically in the Bible but is found as a personal name of one of the five daughters of Zelophehad, *choglah* (Numbers 26:33) and also as a place name *Beth-choglah* (Joshua 15:6).

Proverbial Saying

The partridge brings eggs of other birds and hatches them (*Tanchuma Ki-tetze* 2).

Peacock

The peacock, *tukkiyim* in the plural, is found twice in the Bible: 1 Kings 10:22 and 2 Chronicles 9:21. The *tukkiyim* are among the variety of creatures brought by King Solomon in ships from Tarshish. The name *tukki* is derived either from India where in the old Tamil language it was called *tokai* and in the language Malabar it was *togai* or from Africa where *tukkiym* is a species of monkey. Scholars favor India as the source. Indeed the word is still used in Ceylon, and King Solomon's navy may have visited Southern India or Ceylon and brought back with them the peacock and the gold and ivory mentioned in 2 Kings 10:22 and the Book of Chronicles. Some have suggested that Alexander the Great introduced the bird into Europe.

One writer, enamored of its beauty, portrays it in these words: "His tail . . . surrounded with the colors of the rainbow in most beautiful arrangement . . . and struts about exhibiting the appearance of majesty united with inimitable beauty."[1]

The rabbis anticipated this writer by thousands of years when they proclaimed: "Come and see the peacock, it boasts of 365 different colors and it was created from a drop of the white (of an egg)" (*Tanchuma Tazria* 2).

The Talmud records an incident when a man was offered the head of a peacock in milk and he would not eat it, from which we infer that the peacock was permitted as food but forbidden only with milk (*Shabbat*

130a). Incidentally, the head of the peacock was crowned with a crest and was of delicate texture.

Like the cock, the peacock belongs to the family of partridges, and the rabbis forbade it to be crossbred with the cock: "The cock and the peacock, though they resemble each other are not to be crossbred" (*Yerushalmi, Bava Kamma* 5:8).

In modern Hebrew the word *tukki* is rendered as "parrot," but this cannot be substantiated.

In rabbinic language peacock is rendered as *tavos*; see *Yalkut Esther* 104b where we read of peacocks made of ivory (see also *Pesikta Rabbah* 1).

Pigs (Swine)

Swine is a general name for that class of animals commonly called hogs. The male is designated by the name of boar, the female by that of sow, and the young ones are called pigs. In Hebrew, the term is *chazir* and is found often in the Bible; in Latin it is called *porcus*.

Although the swine is cloven footed, it does not chew the cud and is therefore unclean (Leviticus 2:7). Maimonides comments:

> There is nothing among the forbidden kinds of food whose injurious character is doubted except pork and fat. Pork contains more moisture than necessary for human food, and too much of superfluous matter. The principal reason why the law forbids swine's flesh is to be found in the circumstance that its habits and its food are very dirty and loathsome. It has already been pointed out how emphatically the law enjoins the removal of the sight of loathsome objects, even in the field and in the camp; how much more objectionable is such a sight in towns. However if it were allowed to eat swine's flesh, the streets and houses would be more dirty than any cesspool.[1]

Isaiah brands the use of the *chazir*'s flesh as apostasy (Isaiah 65:4 and 66:17).

The swine is also referred to in Psalms 80:14 as the "swine of the woods" where it swarms along the banks of the Jordan.

The swine symbolizes dirt and mire because of its filthy habits; in Proverbs 11:22 we read, "As a ring of gold in a swine's snout, so is a fair woman that turneth aside from discretion." The above verse contrasts two dissimilar objects that are deliberately juxtaposed to underscore its message. In antiquity the nose ring was a recognized female adornment (Genesis 24:22, Isaiah 3:21). The swine was not only an unclean animal but was also abhorred by Jews. Many who do not observe the Jewish dietary laws nevertheless abstain from eating the flesh of the pig. In this verse the ring is placed in the swine's snout, whereas the expression "fair woman" represents Torah; compare the Mishnah in *Pirkei Avot* 6:2.

There is an interesting reference in the Book of 2 Maccabbees 6:22 in which we are told that Antiochus compelled Jews to eat swine flesh, and Hercanos during his fight against his brother Aristobulos sent a swine over the walls to be offered on the altar. The rabbis then decreed, "Cursed be those who breed swine" (*Sotah 49b*, *Menachot* 64b, *Bava Kamma* 7:7).

Such is the abhorrence of swine that we are warned not even to mention it by name, and the Talmud refers to it as *davar acher* (another thing). It was not only forbidden to breed swine but also to keep them in flocks (*Menachot* 64b; *Bava Kamma* 7:7; *Yerushalmi, Shekalim* 47c).

Incidentally the swine does not discriminate, but eats anything and everything (*Shabbat* 155b), and we understand that its excrement was used by tanners (*Chullin* 17a).

In *Shabbat* 77b we learn that the swine is one of the three animals that grows stronger with age. According to *Kiddushin* 49b, it is the most susceptible to a variety of diseases.

A man had a sow, a she-ass, and a filly. He let the sow eat as much as it wanted, but strictly rationed the ass and filly. Said the filly to the ass, "What is this lunatic doing? To us who do the work of master he gives food by measure, but to the sow, which does nothing, he gives as much as she wants." The ass answered, "The hour will come when you will see her downfall, for they are feeding her up not out of respect for her but to her hurt." When the date of the Roman month for a feast came round they took the sow and stuck it.

Midrash

In rabbinic law a distinction is made between things prohibited for food (*treifah*) and things from which no benefit may be enjoyed, that is, leaven on Pesach. The pig belongs to the former category. Although pig breeding is prohibited, there is no objection to wearing pigskin shoes (even in the synagogue), having a pigskin wallet, or binding religious books in pigskin.

Proverbial Sayings

If a man drinks properly he becomes strong as a lion, whom nothing in the world can withstand. When he drinks to excess, he becomes like a pig that wallows in mire (*Tanchuma, Noah* 13.21b).

The pig said to the clean animals – you should be thankful to us because if we were not here it would not be known that you are clean (*Temurah* 2).

Quail

The quail, *slav*, in Hebrew, is mentioned in Exodus 16:13, Numbers 11:31–32, Psalm 105:40, and Psalm 78:26–30.

In reply to the children of Israel in the wilderness, the Almighty sent down quails. The quail resembles the partridge but it is much smaller. It is a migratory bird, and when it crosses the sea in large numbers it becomes so tired and exhausted that it falls to the ground where it usually nests and becomes an easy prey. The quail is a clean bird, and after the Israelites ate part of the flesh, they spread out the rest to dry in the sun.

That the numbers of quails are vast and enormous was noted by Pliny the Elder. He recorded in his *Natural History* that they sometimes sank the vessels on which they alighted.

The meat of the quail is a delicacy, and as we learn from the Talmud, it is a very fatty bird.

There are four kinds of quail: the thrush, partridge, pheasant, and quail proper, which is like a small bird.

One stuffs it, places it in the oven, and it swells up and becomes so big that it fills the oven. Thereupon one places it on top of twelve loaves of bread, and even the lowest one of them cannot be eaten without some other food in combination (because it is so greasy) (*Yoma* 75b).

Raven

The Hebrew word for "raven," *orev*, is probably connected with *erev* (evening), when darkness descends upon the earth. The raven is usually dark-colored and black; in Song of Songs 5:11, we read, "His locks are curled and black as a raven."

In the Bible the raven is used as a general term for a variety of species, which include crows, rooks, jackdaws, and jays, all of which are found in Israel.

The first time the raven is mentioned in the Bible is in the story of the flood and Noah: "And he [Noah] sent forth a raven, which went forth to and fro, until the waters were dried up from off the earth" (Genesis 8:7). Initially it is difficult to understand why Noah sent the raven on such an important mission, for the raven is an unclean bird and eats carrion. One interpretation informs us that the raven in antiquity was placed aboard every ship because it could smell land. The dove, the second bird sent out, was in a different category; it is a clean bird and feeds on vegetation, a sample of which it held in its mouth. This information enabled Noah to realize that the waters of the flood had abated.

But why the raven? The commentator *Orach Chayyim* suggests that the raven was not sent on any errand, but was driven out of the ark because of his bad behavior. That Noah did not specifically send the raven on a mission is borne out by the interpretation of Rashi on the words, "and it went to and fro." Says Rashi, "And the raven flew in circles round and round the ark and did not go on its errand."

If we follow this interpretation that the raven was expelled from the ark because of its unworthiness, another question arises. Why did God expressly command the ravens to feed Elijah morning and evening when he hid from Ahab in the brook of Cherit (1 Kings 17:4)? It seems strange that the ravens should be rewarded in this manner.

However, it should be remembered that the doctrine of reward and punishment that looms large in the Torah applies not only to man but also to the animal kingdom. The Mezudath David suggests that God deliberately sent to Elijah the ravens, who were cruel by nature, on an errand of mercy to demonstrate to the prophet that he too should not deal harshly with Israel because of their waywardness, but should treat them with kindness and mercy.

The doctrine of reward and punishment is upheld by the *Yalkut* in a remarkable passage. The raven was punished by being ousted from the ark, but he was rewarded when he was attentive and helpful to Adam and Eve who in their desperate plight wept, not knowing how to dispose of the dead body of their son Abel.

In the words of the *Yalkut*, the raven slew one of its mates, dug a hole in the ground, buried it, and covered it with earth. In this manner Adam and Eve performed the first burial. In Jewish law burial of the dead is an act of mercy and truth (*chesed shel emet*). The Almighty rewarded the raven by feeding its young who, because they are born white, are rejected by the parents, who recognize and care for their progeny only when they grow and their plumage is black.

In this manner the Rabbis interpret the verse in Job 38:41 "Who provides for the raven his prey when his young ones cry unto God and wander for lack of food" (*Yalkut Shimoni* 925). To the above we would add the commission granted to the ravens to bring food morning and evening to Elijah, food that was brought from the tables of King Jehosophat.

In Proverbs 30:17 we learn of the cruel habits of the raven who picks out the eye of its victim. Here providence employs the raven to punish one who is guilty of filial impiety: "The eye that mocks at his father and despises to obey his mother, the ravens of the valley shall pick it out." Indeed, the rabbis contrast the raven with the eagle, the former being called cruel (*achzri*) and the latter merciful (*rachmani*).

However, in spite of the apparent cruelty of the parents the Almighty fed the young ravens with the gnats that hovered over the excrement of the parents, and it was these gnats that eventually gave them their black plumage (*Yalkut Shimoni Proverbs* 963).

Rabbi Assi was of an inquiring mind; he once saw a raven making a nest, laying eggs, and hatching fledgelings. He took the young and put them in a new pot and pasted over its top. After three days he opened the top to find out what they were doing, and he discovered that they were secreting excrement, which produced gnats, and the fledglings were devouring them. Rabbi Assi applied to them the verse: "who provides for the raven his food" (*Leviticus Rabbah* 19:1).

Connecting *orev* with *erev*, "evening," the same Midrash provides a novel interpretation of the verse in Job 38:41 "who provides for the raven his prey." According to the Midrash, we learn from Elijah, who studied Torah in the evening. As a reward for having occupied himself early and late with Torah, God appointed for him the ravens to serve him.

In other words, if a man does not show himself as cruel toward his own person, his children, and his household as a raven to its young, he does not succeed in acquiring Torah (*Leviticus Rabbah* 19:1).

Furthermore, the eternal survival of Torah is stressed by Rabbi Alexandri, who said: "If all the nations of the world gathered together to make white one wing of a raven they would not be able to accomplish it. Even so, should all nations of the world gather together to uproot one word of the Torah they would be powerless to accomplish anything" (*Leviticus Rabbah* 19:2).

In the Talmud the raven is portrayed as a champion demanding justice for its species. In a lively argument with Noah the raven said: "Your master hates me and you hate me. Your master hates me since he commanded seven pairs to be taken of the clean creatures, but only two of the unclean (the raven being unclean). You hate me–seeing that you leave the species of which there are seven and send one of which there are only two. Should the angel of heat or of cold smite me would not the world be short of one kind? Or perhaps you desire my mate!" "You evil one," he exclaimed, "even that which is usually permitted me (that of his own wife) has now been forbidden: how much more so that which is always forbidden me!" (*Sanhedrin* 108b).

In another statement in the Talmud we learn that spikes were fixed on the roof of the Temple in Jerusalem to prevent the ravens from nesting there, as it is known that they attack and kill small animals (*Menachot* 107a).

In *Bava Batra* 73b we come across the word *pushkatza* for a large-tailed raven (Jastrow). This bird must have been very strong, for the rabbis observe that it swallowed a snake and perched on a tree. "Imagine how strong was the tree," Rabbi Papa said, "had I not been there I would not have believed it."

In *Bava Batra* 23a ravens are called croakers as we see from the following incident: Rabbi Joseph had some small date trees under which cuppers used to sit and let blood, and ravens used to collect and suck up the blood and fly away to the date trees and damage them. So Rabbi Joseph said to the cuppers: "Take away your croakers from here!"

Shalach in Leviticus 11:17 is usually translated as "cormorant," but Rashi on *Chullin* 63a renders it as "a water raven." The croaking was considered an ill omen (*Shabbat* 67b).

We have seen that ravens reject their young ones when white, but the rabbis confirm that they love each other (*Pesachim* 113b).

As a punishment for copulating in the ark, the spittle ejected from the mouth of the male into that of the female effects conception (*Sanhedrin* 108b).

In *Chagigah* 14a the rabbis interpret the verse in Song of Songs 5:11: "His locks are curled and black as a raven" to refer to the Exodus where God is spoken of as a Man of war (Exodus 15:3). It should be remembered that the Song of Songs is read in the synagogue on Chol Hamoed Pesach.

According to the Mishnah some wealthy people bred domesticated ravens and fed them with a plant bearing beans (*Mishnah Shabbat* 18:1).

Proverbial Saying

See, there's a raven flying past (*urva parach*) (*Chullin* 124b).

Re'em (Wild Ox)

Save me from the lion's mouth, from the horns of the wild oxen do thou answer me" (*Midrash Tehillim*, Psalms 22:22).[1]

While David was tending sheep, he came across the *re'em* asleep in the wilderness, and thinking that it was a mountain he climbed upon it and continued to tend his sheep. The *re'em*, upon waking up, arose, and David astride its horns was lifted as high as the heavens. At that moment David said: "Master of the Universe, if thou bringest me down from the horns of this *re'em* I shall build Thee a temple 100 cubits high, as high as the horns of this *re'em*" (*Yalkut Shimoni* 688).

Some say that David measured the horns of the *re'em* along its length, whereas others maintain that he measured the horns along its circumference. The proof that David made this vow to God is the verse in Psalm 78:69 "And he built his sanctuary like high palaces."

What did the Holy One do for David? He caused a lion to come toward the *re'em*, and when the *re'em* saw the lion he was afraid of it and cringed before it, for the lion is king of all animals and beasts. When David saw the lion he was also afraid of it. Therefore, the Holy One caused a gazelle to come along, and as the lion sprung after it, David descended and went his way. Hence David said: "Save me from the lion's mouth for Thou hast heard me from the horns of the *re'em*."

In Psalm 92:11 it is written: "But my horn shalt thou exalt like the horn of the *re'em*. Like the *re'em* whose horns are so high that it can thrust them to the four corners of the earth."

The *re'em's* strong horn was made into a shofar, and this shofar was also used to announce the Jubilee year (*Rosh Hashanah* 3:5).

The Talmud declares that the young of the *re'em* were taken into the ark. Rabbi Bar Chana said: "I saw a sea *re'em* one day old which was as big as Mount Tabor, which is as big as 4 parsangs. A parsang is a Persian mile equal to 4 English miles" (*Bava Batra* 73b).

The unicorn is found several times in the Bible and is known in Hebrew by the name *re'em*. The unicorn, which is a strong animal, has one horn. It is a land animal and the largest known next to the elephant. Some commentators compare the animal with the rhinoceros.

It has been suggested that the *re'em* is actually an oryx, and it probably refers to an *Ox-bos* (wild ox), as well. Traditional translations of the Bible render *teo* (Deuteronomy 14:5, Isaiah 51:20) as "wild ox," whereas the authorized version and the Jewish Publication Society of America renders it as "antelope."

The strength of the animal is referred to in Numbers 23:22, 24:8: "God, who brought them forth out of Egypt, is for them like the lofty horns of the wild ox." Another reference to the wild ox can be found in Job 39:10, which compares it to the plowing ox: "Canst thou bind the wild ox with his band in the furrow?"

The animal is commonly found in hunting scenes on Assyrian tablets.[2] The Assyrian name is *rimu*.

Proverbial Saying

High (as *re'em*) to the fool is wisdom (Soncino Bible, Proverbs 24:7).

Scorpion

The scorpion is known in Hebrew as *akrav*. Jastrow traces the word to *akev*, "heel," as both the serpent and scorpion sting through the heel. Indeed, the scorpion is often associated with the serpent in scripture. As a creature of the desert it is mentioned once in the Torah: "Who led you through the great and dreadful wilderness wherein were fiery serpents and scorpions" (Deuteronomy 8:15).

Elsewhere in *Tanach* the scorpion is used symbolically. Thus King Rehaboam admonishes the people with these words: "My father chastised you with whips, but I will chastise you with scorpions" (1 Kings 12:11, 2 Chronicles 10:11, 14). Here, scorpions are thorns, the sharp ends of which can sting people as scorpions inject a poison that is very painful. Indeed, the English versions of the Bible render the word *akravim* as "stinging thorns."

In Ezekiel 2:6 the word *scorpion* is used as a metaphor for wicked people. God warns the prophet not to be afraid of the scorpions, who are "a rebellious house." *Targum Jonathan* on this verse interprets *akravim* as real scorpions.

As a place name *akravim* is found in Numbers 34:4 and Joshua 15:3 where the scorpion's pass leads to the northern scope of the wady El-fikreh.

In the Talmud we learn that the spider was used as a cure for the sting of a scorpion (*Shabbat* 77b).

With reference to a statement made by Rabbi Joshua ben Levi that all animals that are dangerous may be killed on the Sabbath, mention is made of the scorpions of Adiabene, a district in Assyria (*Shabbat* 121b). In another statement the rabbis assert that if a child six years of age is bitten by a scorpion, it does not, as a rule, survive (*Ketubot* 50a).

The rabbis record an instance in which a scorpion was carried by a frog across a river and it stung a man who died therefrom.

With reference to this, Samuel pointed out the verse "They stand this day to receive thy judgments for all are thy servants" (Psalm 119:91).

Though the scorpion does not swim it was carried by a frog to fulfill God's command (*Nedarim* 41a).

We learn from the Talmud that Mount Sinai was surrounded by scorpions that "stood up as white asses" (*Bava Batra* 74a).

It appears that the bite of the scorpion is even more dangerous than that of the snake because it repeats it (*Berachot* 9a). The urine of a forty-day-old infant and the gall of the stork were used as antidotes to the poison (*Shabbat* 109b), and warm things are good for a scorpion bite (*Avodah Zarah* 28b). According to *Gittin* 60a the scorpion was employed as a medicament in curing cataract.

The anger of the wise is compared to the sting of a scorpion (*Avot* 2:15).

The iron bit of the horse derives its name from the scorpion because of its shape (*Kelim* 11:5, 12:3).

There are eight to ten species of scorpions in Israel, where they abound.

The scorpion sucks out the sap of its victim and can live without food for months.

Never did a serpent or a scorpion injure anyone in Jerusalem (*Yoma* 21a). For this reason they are common near Jerusalem.

If there is danger of a scorpion, all are permitted to wear shoes (*Yoma* 78b). A visitation of serpents or scorpions is regarded as war (*Yevamot* 114b).

If a man falls into a pit full of serpents and scorpions, evidence may legally be tendered concerning him (to enable his wife to marry again) (*Yevamot* 121a).

If a scorpion is wound round a man's foot he is permitted to break off his prayer because the scorpion will sting (*Berachot* 33a).

Proverbial Saying

Wine is compared to a scorpion. As the scorpion wounds with its tail, so wine wounds at its end (*Numbers Rabbah* 10).

If a scorpion is wound round ... foot he is permitted ... his prayer because the scorpion will sting ... [illegible]

Proverbial Saying

When compared to a scorpion, a ... [illegible]

Serpent

The serpent, also known as the snake, is called in Hebrew, *nachash*, which is derived from the root found only in Pi'el meaning "to practice divination, presage, or observe times." It is frequently found in the Bible, Talmud, and Midrash. Thus Laban craftily says to Jacob: "*Nichashti*–I have observed the signs" (Genesis 30:27).

Rabbi Samson Raphael Hirsch translated *nachash* as "I have superstitious ideas." It is used of Joseph (Genesis 44:5), and Hirsch comments: "Perhaps that is why a snake is called *nachash*." It glides away from that which is in the middle of it and goes about by an indirect course. "He uses it for hidden magic arts."

If we compare the two texts in Deuteronomy 8:15 and Numbers 21:6–9 we learn that in the former text God led the Israelites through the great and dreadful wilderness, wherein were fiery serpents and scorpions, whereas in the latter text God let the poisonous snakes loose and they bit the Israelites because of their grievous sins.

To bring home the lesson of the serpent forcibly, God commanded Moses to make a copper serpent and to suspend it from a pole on high so that the Israelites should raise their eyes heavenward, and the image of the brazen serpent would help them show remorse and so repent of their wrongdoings.

Characterizing the relative merits of his sons, Jacob prophesies and says of Dan that "he will be a serpent on the highway, a rattlesnake on the path that bites the hoof of the horse, and its rider falls backwards" (Genesis 49:17). S. R. Hirsch comments, "When Israel is attacked by military strength, Dan will defend it with snake-like cleverness."

The serpent in Paradise soliloquizes. It argues with itself, saying: "If I go and speak to Adam I know that he will not listen to me, for a man is always hard to be persuaded, as it is said: 'For a man is churlish and evil in his doings' (1 Samuel) but behold I will speak to Eve for I know that she will listen to me, for women listen to all creatures, as it is said: 'She is simple and knows nothing' " (Proverbs 9:13).

In Genesis 3:1–7 we read that the serpent was more subtle than any beast of the field that the Lord had made. In talking to Eve in the Garden of Eden, the snake convinced her to eat of the fruit of the forbidden tree. Consequently, the snake was cursed above all. Because of the great wisdom of the serpent, God's penalty was inflicted upon it proportionate to its wisdom (Ecclesiastes 1:18). The advice of the serpent to Eve brought death to mankind.

The Psalmist rails against unjust judges whose treacherous words are compared to "the venom of a serpent" (Psalm 58:5). This misuse of speech, which is a divine gift, is again underscored by the Psalmist: "They have sharpened their tongues like a serpent" (Psalm 140:4). "Vipers' venom is under their lips" (Psalm 104:4). Here the Psalmist attacks the evils of slander, which, in its insidious propaganda, attacks many innocent people like a sharpened sword.

In a striking warning against drunkenness, Solomon compares habitual drinking to the bite of a poisonous snake: "At the last it (drink) bites like a serpent and stings like a basilisk" (Proverbs 23:32).

In another passage Solomon marvels at "the way of a serpent upon a rock" (Proverbs 30:19). He asks a hypothetical question: "If the serpent is a limbless creature and has no feet, how does it move about?"

The prophet Micah foretells that the nations of the world shall in abasement before God "lick the dust like a serpent" (Micah 7:17), and Isaiah addresses these words to Philistia: "For out of the serpents' root shall come forth a basilisk, and his fruit shall be a fiery flying serpent" (Isaiah 14:29).

Serpents hide behind stone fences and walls. Thus Ecclesiastes records: "He who breaks through a fence, a serpent shall bite him" (*Ecclesiastes* 10:8); Amos writes: "As if a man . . . went into the house, and leaned his hand on the wall, and a serpent bit him" (Amos 5:19); and Isaiah informs us that snakes hatch basilisks' eggs (Isaiah 59:5).

When a man's ways please the Lord, he makes even his enemies eager to give up life (Proverbs 16:7). Rabbi Joshua ben Levi said, "By 'enemy' a snake is meant." He told this story of a man who ground up some garlic and a wild snake came and ate of it and then a tame snake saw it eating. When the people sat down to eat, the tame snake began stirring up dust toward them, but they did not understand what the snake meant, whereupon it threw itself upon the venom-smeared garlic and died (*Genesis Rabbah* 54:1).

The serpent plays a prominent role in rabbinic literature, and the Talmud records some interesting data regarding the serpent: "Can you mark when the hinds do calve? This hind has a narrow womb. When she crouches for delivery, I prepare a serpent which bites her at the opening of the womb, and she is delivered of her offspring; and were it one second too soon or too late, she would die" (*Bava Batra* 16b).

Our Rabbis taught: Four died through the counsel of the serpent, namely, Benjamin son of Jacob, Amram the father of Moses, Jesse the father of David, and Kilab the son of David (*Bava Batra* 17a).

It is written that God created great sea monsters (Genesis 1:21). This refers to leviathan the slant serpent and to leviathan the tortuous serpent, for it is written: "In that day the Lord with his sore (and great and strong) sword will punish them" (*Bava Batra* 74b).

Raba, son of Rabbi Ilai, lectured:

> What is meant by "Moreover the Lord said, Because the daughters of Zion are haughty?" That means that they walked with haughty bearing, and with outstretched necks–they walked heel by toe. And wanton eyes: Walking and mincing: they walked, a tall woman by the side of a short one making a tinkling with their feet. This teaches that they placed myrrh and balsam in their shoes and walked through the market places of Jerusalem, and on coming near to the young men of Israel they kicked

their feet and spurted it on them, thus instilling them with passionate desire like with serpents poison (Isaiah 3:16).

In *Shabbat* 156b we read two stories about snakes. In the first story, Samuel and the astrologer Ablat were sitting while certain people were going to a lake. Ablat predicted that one of the men would die of a snake bite, but Samuel said that he would return if he is an Israelite. The man did return, and upon inspection of his knapsack, a snake was discovered therein, cut up and lying in two pieces. When Samuel asked the man what he had done, the man told of protecting another man from embarrassment at not having bread when everyone pooled their bread and ate it. Samuel told him that he had done a good deed, which "charity delivereth from death."

The second story tells of Rabbi Akiva's daughter whose death from a snake bite was predicted for her wedding day. On her wedding day his daughter took a brooch and stuck it into the wall where by chance it sank into the eye of a serpent. The following morning, when she took the brooch out, the snake came trailing after it. Her father asked what she had done, and she told him of a poor man who had come to their door while everyone was busy at the banquet. She had given her portion to this man. Rabbi Akiva told her that her good deed was a "charity that delivereth from death."

A crushed mosquito is recommended for a serpent's bite. In turn, a serpent is a remedy for an eruption: what is the treatment? One black and one white serpent are brought, boiled to a pulp and rubbed in (*Shabbat* 77b).

According to Rabbi Akiva a serpent can be killed even without a trial (*Sanhedrin* 15b).

How do we know we do not plead on behalf of a *mesit* (inciter to idolatry)? From the story of the ancient serpent (Genesis 3). The serpent had many pleas to put forward, but did not do so. Then why did not God plead on its behalf? Because it offered none itself. What could it have said to justify itself? Where words of the teacher and those of the pupil are contradictory, whose word should be hearkened to, surely the teacher, so Eve should have obeyed the command of God (*Sanhedrin* 29a).

A story is told of a wicked man who slew another. A serpent came and bit the murderer so that he died. But should this man have died through a serpent? It was taught that he who merits burning either falls into the fire or is bitten by a serpent, i.e. the action of the poison is likened unto the inner fire of burning (*Sanhedrin* 37b, *Ketubot* 30b).

Rav said to his son Hiyya: "Do not provoke serpents." Our rabbis taught: "Three must not be provoked–an insignificant gentile, a little snake, and a humble pupil" (*Pesachim* 113a).

Rabbi Simeon b. Manassia said: "Woe for the loss of a great servant. For had not the serpent been cursed, every Israelite would have had two valuable serpents, sending one to the north and one to the south to bring him costly gems, precious stones, and pearls. Moreover, one would have fastened a thong under its tail, with which it would bring forth earth for his garden and waste land." (Before he was cursed the serpent was destined to serve). (*Sanhedrin* 59b).

In Amos 5:19 we read: "As if a man did flee from a lion, and a bear met him; or went into the house, and leaned his hand on the wall, and a serpent bit him" (*Sanhedrin* 98b).

In 2 Kings 18:4 we learn that Hezekiah removed the high places and broke the images, and cut down the groves, and broke in pieces the brazen serpent that Moses had made because the children of Israel were burning incense to it: and he called it *Nechushtan*.

In Isaiah 65:25 we read: "And dust shall be the serpent's food." Rabbi Ammi and Rabbi Assi disputed the meaning of this passage. One said, "Even if the serpent were to eat all the delicacies in the world he would feel therein the taste of dust"; the other said: "Even though he ate all the delicacies of the world his mind would not be at ease until he had eaten dust" (*Yoma* 75a).

Ben Azzai says: "Lie on anything but not on the ground for fear of serpents" (*Berachot* 62b)

Commenting on the verse, "If a serpent bites before it is charmed, then the charmer has no advantage" (Ecclesiastes 10:11), Resh Lakish said, "In the Messianic Age all animals will assemble and come to the serpent and say to him, 'The lion claws his victim and devours him, the wolf tears him and devours him, but as for you what benefit do you derive?' His reply will be: 'The charmer has no advantage' " (*Taanit* 8a).

It has been taught, Rabbi said, "Where it is a case of doing honor we begin at the most distinguished, but where it is a case of censuring we begin at the least important; as it is said: " 'And Moses said to Aaron and to Eleazar and Ithamar; "First the serpent was cursed, and afterwards Eve and only then Adam" ' " (*Taanit* 15b).

There was a serpent in the days of King Shapur (King of Persia) before which thirteen stables of straw were placed, and it swallowed them all (*Nedarim* 25a).

Three liquids may not remain uncovered–water, wine, and milk. The period of time they remain uncovered is the time it takes a serpent to creep out from a place nearby and drink (*Terumot* 8:4).

Rabbi Pinchas related: "A man was standing digging in the valley of Bet Shufre when he saw a certain herb which he gathered and made into a garland for his head. A snake passed by and he struck it and killed it. A certain charmer came along and he halted. 'I am astonished at the person who slew the snake . . . lift the herb off your head.' As he touched the snake his legs fell to pieces" (*Leviticus Rabbah* 22:4).

Rabbi Jannai saw a snake coming. Excitedly he chased it and came back. This creature is designated for the performance of his mission (*Leviticus Rabbah* 22:4).

When Rabbi Eliezer was in a privy, a Roman man made him get up and he sat down in his place. A snake smote the Roman and killed him (*Leviticus Rabbah* 22:4).

In *Lamentations Rabbah* 1:31 we read: Rabbi Johanan ben Zakkai spoke in parables–"If a snake nested in a cask, what is to be done with it?" (The snake is Israel who rebelled against Rome and the cask is Jerusalem.) He answered, "Bring a charmer and charm the snake and leave the cask intact." Pangar said, "Kill the snake and break the cask."

Then they asked, "If a snake nested in a tower (temple), what is to be done with it?" Rabbi Johanan ben Zakkai answered, "Bring a charmer and charm the snake and leave the tower intact." Pangar said, "Kill the snake and burn the tower." Rabbi Johanan said to Pangar, "All neighbors who do harm do it to their neighbors." (The Rabbis pleaded for leniency–Arabs were neighbors of Israel and if a harsh policy were followed by Vespasian, his own people would in due course meet with similar fate.) "Instead of putting a plea for defense you argue for prosecution against us." He replied, "I seek your welfare so long as the Temple exists. The heathen kingdom will attack you but if it is destroyed they will not attack you."

Rabbi Johanan said, "The heart knows whether it is for *akkel* or *akalkalot* (perverseness)" (Your heart knows what your real intention is) (*Lamentations Rabbah 1:31*).

"When a person has been bitten by a snake a piece of rope terrifies him" (They remembered the terrifying experience when they neglected their duty to Joshua so they did not forget their duty when Samuel died) (*Ecclesiastes Rabbah* on 7:1[4]).

Rabbi Isaac said: "It was as if a snake was lying on the crossroads and biting everyone that passed by when a keeper (Jastrow translates this word as dragon) came and sat down facing it. A snake charmer came up, and facing the two of them exclaimed: 'The habit of the snake is to bite. I am surprised at the keeper that he associates with it' " (*Deuteronomy Rabbah* 6:4).

One interpretation of Psalm 1 is that the entire psalm speaks of Adam. Blessed is the man who walks not in counsel of the wicked. Adam said, "If I had not walked in counsel of the serpent how blessed I would have been." Nor stand in the way of sinners. Adam said, "If I had not stood in the way of the serpent how blessed I would have been." Nor sit in the seat of the scornful. Adam said, "If I had not sat in the seat of the serpent, a seat of scorn, how blessed I would have been," and Rabbi Joshua of Siknin related in the name of Rabbi Levi: "The serpent spoke slander against his Creator." The serpent asked Eve, "Why will you not eat of this tree?" And Eve said to him, "God commanded me neither to eat of this tree nor to touch it." What did the serpent do? He picked Eve up and pushed her against the tree, but she did not die, for in truth the

Holy One had only commanded: "You shall not eat of it" (Genesis 2:17). To this divine commandment Eve had added the words "neither shall you touch it" (Genesis 23:3). The serpent then said to her, "Our Creator ate of this tree, then created the world and all that is in it; and if you eat of the tree you will have the power to create a world as He did, for it is said: 'You shall be as God' (Genesis 3:5). But of course every craftsman hates to have a rival in his craft." It follows from this that the serpent was a scoffer.

In the time to come all creatures will be made whole except the serpent, as it is said: "Dust shall continue to be the serpent's food" (Isaiah 65:25).

"Their poison is like the poison of a serpent: they are like the deaf adder that stops her ear" (Psalm 58:5). David said further to them: "Know ye not what The Holy One did to the serpent? He destroyed his feet and his teeth, so that the serpent now eats dust." Even so will The Holy One deal with the maligners, as the next verse says: "Break their teeth, O God, in their mouth" (Psalm 58:7).

That disaster that occurs if one does not follow a qualified leader is illustrated by the fable of the snake whose tail insisted on assuming leadership over the head. Successively, the tail (because of its blindness) led the snake into a pit of water, a fire, and a thicket of thorns (*Deuteronomy Rabbah* 1:10, *Bava Batra* 73a).

Proverbial Saying

No one can live with a serpent in the same basket (*Ketubot* 72a).

Sheep

The name of sheep occurs in some form more than one hundred times in the Bible. Indeed, sheep were used as domestic animals from early times, as early as the days of Abel.

There are many Hebrew words for sheep:

Seh–(Genesis 30:32) is also found in conjunction as *seh kesavim*–"sheep"; *seh ezim*–"goats" (Deuteronomy 14:4)

Keves–"lamb" (Exodus 29:38)

Kesev–(Leviticus 3:7)

Kivsa–"ewe, lamb" (2 Samuel 12:3–4); plural, *kivsat* (Genesis 21:29–30)

Tzohn–a collective noun embracing small cattle, sheep and goats, and flock (Genesis 4.2, Deuteronomy 32:14)

Rahel–"ewe" (Isaiah 53:7, Genesis 31:38)

Ayil–"ram" (Genesis 22:13)

Atud–(Genesis 31:10)

Taleh–(Isaiah 65:25)

Jews were originally rooted in agriculture, and it is therefore not surprising to learn that Moses and David both began their careers as shepherds tending the flock. We are also familiar with the exploits of the patriarch Jacob, who increased his flock many times (Genesis

31:7–9). In this connection it should be noted that the old biblical narrative was re-created in recent years. In 1969 Lady Aldington, wife of the industrialist and former deputy chairman of the Tory party in England, founded the Jacob Sheep Society to preserve a breed described in Genesis 30:39 as striped, spotted, and brindled. These sheep, noted for prolific breeding and for their four horns and black spots, are said to have been descended from ewes rejected by Laban and given to Jacob. Members value the privilege of wearing tweed woven from Jacob's sheep wool.

Of the many references to sheep in the Bible we single out the Paschal lamb, which is central to the festival of Passover. The shank bone, which is exhibited on the Seder table, is reminiscent of the Paschal sacrifice offered on the altar.

Sheep also figure prominently in another festive occasion described in 1 Samuel 25:2–8, where we read of sheepshearing. The shearing of sheep is not considered a mundane affair, but is elevated to the rank of a Yom Tov, a festival.

Sheep are recognized as gentle, docile, and passive; such passive resistance was inherent in the Jewish people, who were continually led to the slaughter as innocent lambs. In Psalm 44:23 we read the phrase "as sheep for the slaughter." Indeed there is ample evidence of this in Jewish history.

In the Talmud we learn that a person should cover himself with the wool of his own sheep and drink the milk of his own sheep and goats. It should also be mentioned that the horn of the sheep was used as a shofar.

Rearing sheep and goats in talmudic times, both in Palestine and in Babylonia, was a profitable occupation. For instance, Rabbi Johanan says, "If one wishes to become wealthy he should devote himself to the rearing of small cattle." It appears that in Babylonia one was prepared to sell his field and invest the proceeds in flocks. Commenting on *vashterot tzonecha* (Deuteronomy 7:13), Rabbi Hisda said that they are so called because they bring wealth to their owners (*Chullin* 84a–b).

In midrashic literature sheep are often found in parable and simile. We begin with Moses, of whom the Torah states that he led the flock to the farthest end of the wilderness (Exodus 3:1). Commenting on this verse the Midrash tells a delightful story about how Moses observed that one of the sheep was straying and going to the waters of the brook to slake its thirst. Moses addressed himself to the dumb animal in these words: "Had I known that you were thirsty I would have taken you in my arms and carried you to the brook." It is also interesting to note that during the forty years that Moses acted as a shepherd not one sheep was attacked by wild beasts nor was a single sheep lost. The Almighty rewarded Moses for his kindness and consideration by appointing him leader and liberator of the Children of Israel.

Commenting on the verses in Isaiah 7:1 and 9:11, the Midrash adds "If there will be no kids there will be no wethers, no wethers, no flock, no flock, no shepherd, and without a shepherd there will be no world." In other words, if there are no children there will be no young students, no students, no scholars, no scholars, no sages, no sages, no prophets, and when there are no prophets God will not cause His divine presence to rest upon people. Hence, it is written in Isaiah 8:16, "Bind up the testimony, seal the law among my disciples" (*Genesis Rabbah* 42:3).

In another passage we compare the timidity of Israel to that of sheep. In *Leviticus Rabbah* Israel is compared to sheep. As with a lamb, when it is hurt on its head or any other limb, all its limbs feel it (this is true of all animals, but particularly of sheep, because they are small and weak), even so is it with Israel; if only one sins, all feel it.

According to the Midrash, Titus once looked on a marketplace and saw many heathens milling around and a Jew among them, and no one said a word against him. He was amazed because he knew that none of them loved Jews. If they were to trample him down, no one would have known. He told this to Akiva saying he had seen something remarkable – among thousands of lions and wolves he saw a little lamb. Akiva understood what he meant and answered they did nothing to the lamb because they were afraid of the shepherd. Titus was amazed by the answer.

A story is told of a poor European Jew who lived in a one-room hut and owned just one little sheep. One day he heard that Cossacks were coming to take his little sheep away, and he loaded his gun. When the would-be robbers surrounded his hut, he ran furiously from window to window shooting at his enemies. In hurrying back and forth in his little room, he began to stumble over the little sheep that he was protecting. He was soon terribly annoyed by the sheep's interference, and without thinking, he opened the door and pushed out the sheep and then returned to the task of fighting the surrounding Cossacks.

It is told that the Baal Shem Tov was once obliged to celebrate the *Shabbat* in the open fields. A herd of sheep was grazing not far off. When he spoke the blessing to greet the approaching *Shabbat* bride, the sheep stood on their hind legs and remained in this position, turning toward the master until he had ended his prayer.

Proverbial Sayings

If no vineyard, why a fence; if no sheep why a shepherd? (*Mechilta* on Exodus 12:1).

When the shepherd is on the right path, the sheep too are on the right path (*Pirke d'R. Eliezer*).

Great is the sheep (Israel) that lives among seventy wolves (nations) (*Tanchuma Toledot* 5).

Sparrow

The Hebrew name for sparrow is *tzippor deror*, which is translated as "bird of freedom," and it is found in Psalm 84:4: "Yea, the sparrow hath found a house, and the swallow a nest for herself." From this verse we learn that the sparrow builds its nest near human habitation (see also Psalm 102:8).

The name of the bird is derived from the fact that it lives in houses as well as in fields (*Betzah* 24a). Though it is found near human dwellings it refuses to be domesticated. The bird is found in Israel in large numbers at all times of the year.

The sparrow is a clean bird and is therefore fit for Jewish consumption (*Chullin* 140a).

The Talmud calls the sparrow *tzaporta*: "The quail looks like a *tzaporta*" (*Yoma* 75b). It says that the sparrow resembles the bird, the blood of which purified the leper (Leviticus 14:4): "He should bring two house sparrows" (*Nega'im* 14:1, *Sotah* 16b).

In *Lamentations Rabbah*, proem 20, reference is made to Psalm 102:8: "As a sparrow is driven from roof to roof, fence to fence, tree to tree, bush to bush . . . so had I to wander in Israel for forty years in exile."

Midrash Tehillim on Psalm 84:4 reads, "The children of Israel said: 'How long shall our enemies be allowed to hate us and say as a sparrow' "[1] (compare Proverbs 26:2). Our enemies say to God: "Flee as a sparrow to your mountain" (Psalm 11.1). They do not say flee as a dove, but flee as a sparrow. A dove, even when her fledglings are taken away, returns to her nest, as Scripture says, "like a silly dove without understanding" (Hosea 7:11). Yet, this is not so with the sparrow. True, she hatches her chicks in her nest, but when her fledglings are taken away from her she does not go back to the nest.

Proverbial Saying

One sparrow doesn't bring the spring season (*Bat Shlomo* 31, *Hegdarim* 8).

Spider

The Hebrew word *achavish*, "spider," reminds us of the root *chavish*, "to subdue." There is probably no real connection between the two words, however, although the spider can subdue and even kill any creature that is entangled within its web.

The spider is mentioned three times in the Bible. In Isaiah 59:5, 6 we read of "the wicked who hatch basilisk's eggs and weave the spider's web"; the basilisk's eggs probably mean in this passage the destructive ways of man. In the Book of Job 8:14 we read: "whose confidence is gossamer, and whose trust is a spider's web." As the spider's web is fragile and tender, so are the hypocrites and the wicked who put their trust in the spider's web. In another passage (Psalm 140:4) we find the word *akhshub*, which is a variant of *achavish* and is considered to be a poisonous creature (see Rashi).

Some authorities suggest that *semamit* (Proverbs 30:28) is a spider, but the majority opinion favors its translation as lizard, rather than spider.

In *Genesis Rabbah* 66:7 and Proverbs 30:28 this hypothetical question is asked: in virtue of what merit does the spider grasp? In the merit of those hands whereof it is written, "and he also made savory foods." The spider symbolizes Esau–Rome, which has grasped dominion as a reward for respect shown by Esau to his father by preparing savory foods for him with his own hands. The verb is used actively. The spider seizes with his own hands.

The Talmud refers to the oft-quoted saying that "passion is at first like a spider's web and afterward it grows as strong as the ropes of a wagon" (*Sukkah* 52a; *Sanhedrin* 99b).

Each spider family has its own particular pattern for web spinning. Scientists can distinguish to which family a spider belongs by studying the web. The baby spider can also weave its own design by learning from its mother.

There is a fascinating story in which King David, who was fleeing from Saul's wrath, is saved by a spider's web. It is told in the following Midrash:

It once happened that King David while sitting in his garden noticed that a hornet was devouring a spider. Said David to the Almighty: "Master of the Universe, for what purpose did you create these two creatures? The hornet destroys the spider and has no pleasure from it; the spider weaves a whole year and is devoid of any clothing." The Almighty responded to David in these words: "David, you seemingly mock my creatures, but the day will come when you will need them."

When David hid himself in the cave to escape King Saul's anger, the Almighty arranged that a spider should weave its web at the entrance of the cave, and in this manner the entrance was closed. When Saul appeared and noticed that the entrance was closed, he surmised that no one was in the cave, arguing that if anyone had entered, the web would be torn apart.

When David left the cave and saw the spider, he almost embraced him. "Blessed be He who created you and blessed be thou."

When David came upon Saul asleep in the midst of a defensive circle of wagons and Abner was guarding him, David crept in and lay between his feet while removing a cruse of water. When he tried to extricate himself from the feet of Abner, the latter stretched out his legs, which had the effect of pinning David down. In his plight David beseeched the Almighty, crying out, "My God, why hast thou forsaken me!" Immediately the Almighty sent a hornet that stung Abner's feet, allowing David to escape. Whereupon David praised the Lord and said,

"How manifold are thy works O Lord, all Thy deeds are pleasant and mighty" (Alef Beth of Ben Sira).

The *Yalkut Shimoni* testifies that the spider is the creature most hated of man (*Yalkut Shimoni* 140c).

In post-talmudic literature Bachya ibn Pakuda writes in his *Duties of the Heart* that "as the cobweb obstructs the light of the sun, so does passion the light of reason."

Stork

The stork, which is an unclean bird, is referred to in Leviticus 11:19 and Deuteronomy 14:18. It is known by the name of *hassidah*, which means kindness, and the rabbis state that it deals kindly with its companions (*Chullin* 63a). The question has therefore been posed: if it shows kindness why should the stork be unfit for Jewish consumption? The reply is that kindness must not be limited to one's friends, but should rather be universal. It is interesting to note that the stork was regarded among all ancient Oriental races as a symbol of devoted maternal and filial affection. It is said to be unrivaled among birds for an affectionate and amiable disposition, and in Egypt it was looked upon as the symbol of a dutiful child.[1]

From this Egyptian symbolism may come the origin of the folklore that the stork delivers the baby. In this connection it is worth noting that the number of babies delivered was recently proved by a professor of statistics to be directly dependent on the stork population.[2]

The stork is a migratory bird, as we learn from Jeremiah 8:7: "Yea, the stork in the heaven knoweth her appointed times." From Psalm 104:17, "As for the stork, the fir trees are its house," we may assume that the stork nested in Israel in the biblical era. From Zechariah 5:9, "and the wind was in their wings," we learn that the stork has large flapping wings that produce a mighty sound.

Tortoise

The tortoise is mentioned only once in the Bible, in Leviticus 11:29. The Hebrew word *tzav* is derived from a root meaning "to swell, or to be arched," as the tortoise has an arch over its body. The same word *tzav* is also used for the arched covering of a wagon (Numbers 7:3, Isaiah 66:20).

There are several kinds of tortoise: the sea tortoise, the fresh-water tortoise, and the land tortoise. The sea tortoise is usually called a turtle. The Midrash includes the salamander in the definition, "the *tzav* after its kind" (*Sifrei Shemini* 6:5).

Some commentators define the tortoise as the lizard. The Talmud points out that the lizard mates with the snake, and the offspring is the *arod* (*Chullin* 127a), a species of poisonous reptile.

Rashi finds a resemblance between the *tzav* and the frog (*Kiddushin* 80a). This interpretation has led some commentators to compare the *tzav* with the toad.

Viper

There are three different types of viper: the Palestine viper, the carpet viper, and the Cerastes viper.

The Palestine viper is known as the *tzefah tzifoni*. According to Mishnah *Avot* 5:8, "No harm will befall anyone on account of an attack by snakes or scorpions in Jerusalem." The bite of the Palestine viper is different from that of nonpoisonous snakes. As we learn from Proverbs 23:32, "it biteth like a serpent, and stingeth like a viper." The Palestine viper lays eggs that immediately open and produce the offspring. This is evident from Isaiah 59:5, where we read: "They hatch vipers' eggs . . . he that eateth of their eggs dieth" (see also Isaiah 14:29, 11:8; Jeremiah 8:17).

The carpet viper, known in Hebrew as *eph'eh*, is the most dangerous of all the poisonous snakes found in Israel. This viper frequents the southern parts of Israel, as we read in Isaiah 30:6, "the South, through the land of trouble and anguish, the carpet viper, and the fiery flying serpent." The Midrash refers to the carpet viper as *ekhes*, thus equating it with *eph'eh* (*Mechiltah Beshalach*, chapter 2; *Tanchuma Beshalach*, chapter 3). In Proverbs 7:22 it is referred to as *eches*. The pink tongue of the carpet viper can kill: "The carpet viper's tongue shall slay him" (Job 20:16). The Midrash further refers to the viper's longevity: "*Eph'eh* gives birth at the age of seventy" (*Bechorot* 8a).

The Cerastes viper is known in Hebrew as *shephiphon*, which is the onomatopoeic imitation of the sound it produces through the "scraping" of its scales on the ground. The characteristics of this viper are described in the Blessings of Jacob (Genesis 49:17). Snake poison lives on even after the death of a snake itself, as we read in *Genesis Rabbah* 98:14.

Proverbial Saying

Balaam was lame in one foot, for we read (Numbers 13:3) *shefi*. Samson was lame in both feet, for it says (Genesis 49:17) *shephiphon* (sliding) on the road (*Sotah* 10a).

Weasel

The weasel is found only once in the Bible, where it is referred to as *choled* (Leviticus 11:29). The name *choled* signifies earth, the dust of which is reddish; and with a word (*chaluda*) used by the rabbis for rust, which is usually reddish.

In Psalm 49:2, the word *choled* is translated as "world." The rabbis use rare words to teach lessons from nature: "All the inhabitants of the world are like the weasel, *choldah*."

An interesting story is recorded by the rabbis:

Once Rabbi Judah the Prince sat and taught the law before an assembly of Babylonian Jews in Sepphoris, and a calf passed before him. It came and sought to conceal itself, and began to moo, as if to say "save me." Then he said, "What can I do for you? For this lot (i.e., to be slaughtered) you were created." Hence Rabbi Judah suffered a toothache for thirteen years. . . . After that a reptile (perhaps a weasel) ran past his daughter, and she wanted to kill it. He said to her, "Let it be, for it is written, 'His mercies are over all His works.' " So it was said in heaven, "Because he had pity, pity shall be shown to him," and his toothache ceased (*Genesis Rabbah* 33:3, *Bava Metzia* 85a).

There is an allusion to a fascinating story of the weasel and the well in the Talmud (*Taanit* 8a). The story is introduced by these words: "R. Hanina said, 'Come and learn the greatness of men of faith from the story of the weasel and the well.' "

According to a note in Malter's edition of *Taanit*, we learn that the story here alluded to is not found anywhere in the Talmud or the Midrashim, although Rashi says that it was quite popular.[1] The only source where it is given in full is the *Aruch* of Nathan b. Jehiel of Rome (11th century). With some omissions, it runs as follows:

A well-dressed, pretty girl who was going home lost her way and, after wandering about in the field for some time and becoming exhausted, came near a well that was provided with a rope and bucket. To slake her thirst, she slid down on the rope, but was unable to pull herself up again. Her cries attracted the attention of a young man who happened to pass by. He offered her help if she promised to marry him, which she did, whereupon he pulled her up. As there were no witnesses to testify to the pact of betrothal, as required by Jewish law, the girl suggested that the well and a weasel that was seen near it should be invoked as witnesses. This done, the couple parted. The young man however, broke his promise and married another girl, who bore him a son. At the age of three months, the boy was choked to death by a weasel. Later a second boy was born to the couple, and he fell into a well and was drowned. The mother, becoming alarmed by the peculiar nature of these accidents, asked her husband for some explanation. He told her the story of his former breach of promise, whereupon she demanded a divorce and obtained it. During all that time, his former fiancée was waiting for his appearance, discouraging all her suitors by feigning epileptic fits. When the young man, after a long search, finally found her, she did not recognize him and tried to deter him in the same way. When he mentioned to her the weasel and the well, she at once accepted him. They were married and lived happily together, their marriage being blessed with children (compare Rashi and Tosafot, ad locum, *Midrash HaGadol*, *Sefer Hamasiot*).

Saadia Gaon and others render *choled* as "mole." In the Talmud the ordinary species of weasel are mentioned under the names of *choldah* and *charchoshtah* (*Pesachim* 9a; *Sanhedrin* 105a). References to the weasel are also found in *Pesachim* 8b and 118b, *Niddah* 15b, and Rashi on *Sukkah* 20b. It can even attack corpses (*Shabbat* 151b). In *Chullin*

52b we learn that it is especially dangerous because it attacks domestic fowl. Its bent and pointed teeth can pierce the skulls of hens (*Chullin* 56a; compare Rashi on Deuteronomy 32:5).

Proverbial Saying

Weasel and cat (when at peace with each other) had a feast on the fat of the luckless (*Sanhedrin* 105a).

Wolf

The wolf, *zev*, is characterized in the Bible as a most ferocious and cruel animal (Genesis 49:27): "Benjamin is a wolf that raveneth; in the morning he devoureth the prey, and at even he divideth the spoil" (compare also Isaiah 11:6, 65:25; Jeremiah 5:6). "Those who persecute the truth are often compared to wolves" (Ezekiel 22:27).

In *Habakkuk* 1:8, the horses of the Chaldean are said to be more fierce than the evening wolves. From this we learn that wolves were very fierce and made their ravages in the evening. In Zephaniah 3:3 the princes of Judah are compared to evening wolves.

According to the Talmud the wolf resembles in external appearance the dog, with which it can copulate (*Berachot* 9b; *Genesis Rabbah* 31:6). However, its howl is even louder and more terrifying than that of the dog. It is particularly hostile to the sheep and vents its spleen against the he-goats (*Shabbat* 53b).

The rabbis refer to the dangerous bite of the wolf (*Zevahim* 4b): "and when hungry it even attacks man" (*Taanit* 19a). We also learn that the wolf can be tamed (*Sanhedrin* 15b), and seemingly it is compared to the otter (*Sukkah* 56b; *Genesis Rabbah* 112:3).

Regarding fables of which the wolf is the subject, see Rashi on *Sanhedrin* 39a and 105a.

The rabbis inform us that once the wolves killed more than 300 sheep (*Betzah* 60a). This statement may have given rise to the saying found in the Midrash that Israel, who is portrayed as a lamb (Jeremiah 50:17), is hotly pursued and relentlessly attacked by the bloodthirsty and rapacious wolves of mankind. Compare the following Midrash, "Hadrian said to Rabbi Joshua: 'Great indeed must be the lamb, Israel, that can exist among seventy wolves.' He replied; 'Great is the Shepherd who rescues and protects her' " (*Tanchuma*, *Toledot* 5).

There are many references to wolves in the Midrash. Thus it is reported that Rabbi Nathan said, "Even a dog or wolf is answerable." This follows the saying "None becomes answerable for injury done to man save another man like himself" (this excludes animals as per Rabbi Nathan) (*Genesis Rabbah* 26:6).

In another Midrash we read: "Who was likened unto a beast of the field? Benjamin. Jacob blessed Benjamin in allusion to the Empire of Media, the former being likened to a wolf, and the latter likened to a wolf. By whose hand will the Empire of Media fall? By the hand of Mordechai, descended from Benjamin."

A story similar in vein to the above is found in the collection of fables attributed to the Russian writer Ivan Andrevich Krylov (1768–1844). He relates that on a hot day a gentle lamb approached a stream of water to quench his thirst. A hungry wolf appeared and marked the helpless lamb for his prey. Wishing to justify his act the wolf reprimanded the lamb. "How dare you with your dirty snout spoil my clear water, causing a buildup of mud and slime?" The poor lamb attempted to convince the wolf that one could not possibly contaminate this running stream. The wolf was not impressed and claimed that if this lamb had not done it that year, it was guilty of having spoiled the stream of water two years ago. The lamb rejected this accusation, pointing out that he was less than a year old. The wolf claimed that if this lamb was innocent, then his brother was guilty. Again, the lamb defended itself by asserting that he had no brother. In reply to the wolf's accusation that his father or grandfather was responsible for this sin, the lamb justifiably claimed that he should not be held responsible for the sins of his antecedents. Finally, the hungry wolf said to the lamb, "Your guilt consists in this, I want to eat you up." Hungry nations, like wolves, want to eat of Israel.

We conclude this section on a happy note: We also learn from *Ecclesiastes Rabbah* 9:1 that "a time will come when the wolf will have fleece of fine wool and the dog a coat of ermine (to provide garments for the righteous)."

Proverbial Sayings

The attack by one wolf is not considered an accident relieving it from responsibility (*Bava Metzia* 93b).

The wolf sheds its hair but not its nature (*Mivchar Hapnimim* 27:4).

The wolf won't be stopped by a fence from coming to the herd (*Keter Malchut*).

Worm

There are two Hebrew words that signify the worm: *tole'ah* and *rimmah*.

There are several references to the worm in the Bible. For instance, the Psalmist says, "But I am a worm and no man" (Psalm 22:7; Job 25:6.) This verse testifies that David, in his humility, considered that he was not worthy to be called a man.

In another verse, Israel is compared to the worm: "Fear not, thou worm Jacob, and ye men of Israel" (Isaiah 41:14). The worm here symbolizes lowliness. In other verses, the worm, combined with fire, connotes eternal pain (Isaiah 66:24). We know that the worm feeds on corpses: "the worm is spread under thee, and worms cover thee" (Isaiah 14:11) and "look upon the carcasses of the men that have transgressed against me: for their worm shall not die" (Isaiah 66:24).

The rabbis interpret that the verse "all that go on their bellies" (Leviticus 11:42) includes the earthworm and all its species (*Chullin* 67b).

As the raven is cruel toward its young, Providence assists them by causing maggots to arise from their excrement (Rashi in *Bava Batra* 8a). According to *Sanhedrin* 108b, Noah fed the chameleons in the ark with the worms that arose from rotten bran. The rabbis inform us that there are a variety of worms in the liver and the belly (*Shabbat* 109b). A cure for worms is mentioned by the sages, who suggest that the milk of an ass mixed with the bay leaves, or bread and salt taken from fresh water, should be eaten before breakfast (*Gittin* 69b; *Bava Metzia* 107b). Garlic is a cure for worms in the great intestine (*Bechorot* 82b), whereas the tapeworm is driven out by the raw meal of barley or by hyssop (*Berachot* 36a, *Shabbat* 109b). *Moranah* is the name of a worm that lodges between the prepuce and glans penis and is removed by circumcision, so that even gentiles submitted to the operation (*Avodah Zarah* 26b). Out of the mouths of false spies and heretics came forth worms (*Sotah* 35a, *Yoma* 1:5; compare also *Yoma* 19b and *Bava Metzia* 84b).

In one midrashic passage (*Midrash Tehillim*, Psalm 22) the rabbis discuss the mouth of the worm: "But I am a worm."[1] Like a worm whose only resource is its mouth, so the children of Israel have no resource other than the prayers of their mouths. Like a worm that roots out a tree with its mouth, so the children of Israel with the prayers of their mouths root out the evil decrees that the hostile nations of the earth devise against them (see Isaiah 41:14–15).

In the famous work *Duties of the Heart* we read: "Observe a silken cord. How strong it becomes when it is doubled many times, though its origin is the weakest of things – a worm's mucus. Note also how a ship's thick cable, when it has been used for a long time, gradually wears out and breaks, so with grave and light transgressions."[2]

The rabbis always stress that every creature has its purpose, and this is corroborated by scientific thought. Experiments have been conducted using the worm for garbage eradication, as it was found that a ton of worms can devour up to 500 kilograms of garbage per day. Further experiments were initiated to use the worm as cattle feed and human food.

Without the earthworm all vegetation would perish and no life could exist on earth.[3]

Proverbial Sayings

As the worm has (for his defense) nothing but its mouth, so Israel has nothing but the prayer of his mouth (Midrash on Psalm 22:7).

A worm is as painful to the dead body as a needle in sound flesh (*Berachot* 18b).

The prospects of man are worms (*Avot* 4:4).

Ziz

Ziz is translated by the B. D. B. Hebrew Lexicon as "a beast, a moving thing," and Rashi, on Psalm 50:11, comments that it moves from place to place. Ibn Ezra, on Psalm 80:14, calls it a bird. The Targum, on Psalm 50:11, calls it a wild cock.

The *ziz* is a reptile and comes under the group of reptiles mentioned in Leviticus 11:41–42. Some commentators suggest that *ziz sadai* in Psalm 50:11 also refers to reptiles. The Soncino Bible translates *ziz sadai* as "the wild beasts of the field." Compare Psalm 80:14 where the same phrase is used, and the following note is included: "As wild animals gnaw the grapes and tender shoots, so the land has been desolated by enemies."

Ziz ruled over the birds as leviathan, king of the fish, ruled over sea life. He was reputed to be as big as the leviathan, and tradition says that when his ankles rested on earth, his head reached the sky and his wings were so huge that they obscured the sun, but they also acted as a protection to earth against winds. The eggs of the *ziz* are truly enormous, but the mother bird is always careful to let her eggs slide gently into the nest. Once, an egg of the *ziz* fell to the ground with a mighty crash and broke into smithereens. By its fall it crushed 300 cedar trees, and the liquid from the egg flooded 60 cities. We read in the Talmud that Rabbah ben Bar Hana related: "Once we traveled on board a ship and we saw a bird standing up to its ankles in the water." Rabbi Ashi said that this was *ziz sadai* (*Bava Batra* 73b).

In *Genesis Rabbah* 19:4 we read that the *ziz* is a clean bird (compare Psalm 50:11) and may be eaten. *Leviticus Rabbah* calls the *ziz* a clean fowl and asks, "Why is it called *ziz*?–because it possesses many kinds of tastes–the taste of this and of that." *Numbers Rabbah* 21:18 refers to the *ziz* as a "fowl of heaven" so huge that it darkened the sun (metaphorically). In *Genesis Rabbah* 71:5 we read: Out of tribes and exalted priestly royalty Aaron's rod blossomed. The priesthood rod was entrusted to Judah, who was considered royalty.

Notes

Preface

1. Henry Knobil, "Some Thoughts on the Law of *Shatnes*," in *Lev Avot on Pirkei Avot* (Jerusalem: Toperoff, 1984), p. 75.

Sabbath and Festival

1. Israel Abrahams, *The Book of Delight* (Philadelphia: Jewish Publication Society, 1912).

Animal Welfare

1. Samson Raphael Hirsch, *The Pentateuch* (New York: Judaica Press, 1971), Genesis 33:13.
2. Ibid., Leviticus 22:28.
3. J. H. Hertz, *The Pentateuch* (London: Soncino, 1960), Leviticus 22:28, p. 518.
4. Ibid., Deuteronomy 22:4, p. 843.
5. Ibid., Deuteronomy 22:10, p. 844.

Post-talmudic Literature

1. *Al Khazari*, ed. Hartwig Hirschfeld (London: Cailingold, 1931), p. 131.
2. Joseph Albo, *Ikkarim*, vol. 4, pt. 1, ed. Isaac Husik (Philadelphia: Jewish Publication Society, 1929), pp. 90–92.

3. Ibid. vol. 1, p. 73.

4. Maharal, *Kol Kitvei Hamaharal,* vol. 2., ed. Abraham Kariv (Jerusalem: Mosad Harav Kook, 1948), p. 119.

5. Noah Rosenbloom, *Studies in Torah—Luzzatto's Ethico-Psychological Interpretation of Judaism* (New York: Yeshiva University Press, 1965), p. 79.

6. Asher Asher, *The Jewish Rite of Circumcision* (London: Shapira Valentine, 1873), p. 102.

7. Arthur Marmorstein, *Studies in Jewish History* (London: Oxford University Press, 1950), p. 32.

8. Leo Jung, *Guardians of Our Heritage* (New York: Bloch, 1958), p. 47.

9. Ibid., p. 49.

10. Ibid., p. 86.

11. Ibid., p. 315.

12. Ibid., p. 492.

13. J. Wohlgemuth, "Consideration for the Animal in Judaism," *Tradition* (1967): 326–327.

14. Solomon Goldman, *The Book of Human Destiny* (New York: Harper & Brothers, 1949).

Perek Shirah

1. Joseph Albo, *Ikkarim,* vol. 2, ed. Issac Husik (Philadelphia: Jewish Publication Society, 1929), p. 12.

Ant

1. Arieh Shoshan, *Animals in Jewish Literature* (Rehovot: Shoshanim, 1971), p. 243.

2. Marcus Jastrow, *Dictionary of Targum, Talmud, and Midrash* (New York: Pardes, 1950), p. 913.

3. Adolf Jellinek, *Beth Hamidrash*, vol. 5 (Vienna, 1878), p. 22.

4. Joseph Albo, *Ikkarim*, vol. 3, ed. Isaac Husik (Philadelphia: Jewish Publication Society, 1929), p. 10.

Ass

1. Maimonides, *The Guide for the Perplexed*, ed. M. Friedlander (London: Routledge, 1928), p. 288.

2. Morris Seale, *The Desert Bible* (London: Wiedenfeld & Nicholson, 1974), p. 88.

Bear

1. Brown, F., Driver, S. R., and Briggs, C. A., eds., *Hebrew and English Lexicon* (Oxford: Clarendon, 1929).

2. Ludwig Lewysohn, *Die Zoologie des Talmuds* (Frankfurt: J. Baer, 1858).

Bee

1. J. Newman, *Agricultural Life of Jews in Babylon* (London: Oxford University Press, 1932), p. 136.

2. Maimonides, *The Guide for the Perplexed*, ed. M. Friedlander (London: Routledge, 1928).

3. Joseph Albo, *Ikkarim*, vol. 1, ed. Isaac Husik (Philadelphia: Jewish Publication Society, 1929), p. 40.

4. Avigdor Miller, *Rejoice O Youth* (New York: Miller, 1962), p. 16.

Birds

1. Yehuda Feliks, *The Animal World of the Bible* (Tel Aviv: Sinai, 1962), p. 87.

2. *The Disputation* (Manchester, England: Scholarly Publications, 1972), p. 41.

3. Maurice H. Farbridge, *Biblical and Semitic Symbolism* (London: Kegan Paul, 1923), p. 59.

Camel

1. Joseph Albo, *Ikkarim*, ed. Isaac Husik (Philadelphia: Jewish Publication Society, 1929) 4:1, p. 92.

Cat

1. Arieh Shoshan, *Animals in Jewish Literature* (Rehovot: Shoshanim, 1971), p. 79.
2. Avigdor Miller, *Rejoice O Youth* (New York: Miller, 1962), p. 13.
3. Joshua Trachtenberg, *The Devil and the Jews* (Cleveland: Meridian, 1961), p. 26.
4. Ibid., p. 218.

Cock and Hen

1. Israel Abrahams, *Companion to the Daily Prayer Book* (London: Eyre & Spottiswoode, 1932), p. xvi.
2. Arieh Shoshan, *Animals in Jewish Literature* (Rehovot: Shoshanim, 1971), p. 83.
3. Solomon Schechter, *Studies in Judaism*, 2d. series (Philadelphia: Jewish Publication Society, 1908), p. 175.

Dog

1. Louis Ginzberg, *Legends of the Jews*, vol. 6 (Philadelphia: Jewish Publication Society, 1946), p. 368.
2. Bachya ben Joseph, *Duties of the Heart*, vol. 2, trans. Moses Hyamson (Jerusalem: Boys Town, 1965), p. 99.

Eagle

1. H. Tristam, *Natural History of the Bible* (London: Society for Promotion of Knowledge, 1898).

2. R. Joseph Ben Judah Ibn Aknin (1160–1237) in his commentary on *Mishnah Avot.*

Elephant

1. S. Singer, ed., *Authorised Daily Prayer Book* (London: Eyre & Spottiswoode, 1950), p. 389.

Fish

1. C. Montefiore and H. Loewe, *A Rabbinic Anthology* (Philadelphia: Jewish Publication Society, 1960), p. xcvi.
2. Ibid., p. xcii.

Gazelle

1. Maurice H. Farbridge, *Biblical and Semitic Symbolism* (London: Kegan Paul, 1923), p. 70.
2. W. Robertson Smith, *Kinship and Marriage in Early Arabia* (London: A. C. Black, 1903), p. 227.
3. W. Robertson Smith, *Lectures on the Religion of the Semites* (1889, 1894), p. 444.
4. Yehuda Feliks, *The Animal World of the Bible* (Tel Aviv: Sinai, 1962).

Goat

1. Maimonides, *The Guide for the Perplexed*, ed. M. Friedlander (London: Routledge, 1928), p. 364.
2. Joshua Trachtenberg, *The Devil and the Jews* (Cleveland: Meridian, 1961), p. 47.

Hornet

1. J. Garstang, *The Fundamentals of Bible History* (London: Constable, 1931).
2. Samuel Driver, *International Critical Commentary on the Holy Scriptures* (Edinburgh: Clark, 1895).
3. Yehuda Feliks, *The Animal World of the Bible* (Tel Aviv: Sinai, 1962).

Horse

1. Solomon Thieberger, *King Solomon* (Oxford: East and West, 1947), p. 156.
2. Sixty respirations are the talmudic definition of a "casual sleep" or a nap.

Horse Leech

1. Jonathan Fisher, *Scripture Animals* (New York: Weathervane, 1977) p. 150.

Leopard

1. R. Joseph Ben Judah Ibn Aknin (1106–1237) in his commentary on *Mishnah Avot.*

Leviathan

1. Maurice H. Farbridge, *Biblical and Semitic Symbolism* (London: Kegan Paul, 1923), p. 75.

Monkey or Ape

1. Avigdor Miller, *Sing You Righteous* (New York: Miller, 1971), p. 122.

Mosquito (Gnat)

1. *The Disputation* (Manchester, England: Scholarly Publications, 1972), p. 22.

Moth

1. Yehuda Feliks, *The Animal World of the Bible* (Tel Aviv: Sinai, 1962).

Mouse

1. Moshe Kleinman, *Or Yesharim* (Lodz: Mesorak, 1934), p. 79.

Ostrich

1. Yehuda Feliks, *The Animal World of the Bible* (Tel Aviv: Sinai, 1962).
2. H. Tristram, *Natural History of the Bible* (London: Society for Promotion of Knowledge, 1898).

Owl

1. Jonathan Fisher, *Scripture Animals* (New York: Weathervane, 1977).
2. Yehuda Feliks, *The Animal World of the Bible* (Tel Aviv: Sinai, 1962).
3. Ibid.
4. Ibid.
5. Walter Ferguson, *Living Animals of the Bible* (New York: Scribner's, n.d.), p. 48.
6. Feliks, *Animal World.*

Partridge

1. Jonathan Fisher, *Scripture Animals* (New York: Weathervane, 1977), p. 209.

Peacock

1. Jonathan Fisher, *Scripture Animals* (New York: Weathervane, 1977), p. 213.

Pigs (Swine)

1. Maimonides, *The Guide for the Perplexed*, ed. M. Friedlander (London: Routledge, 1928), pp. 370–371.

Re'em (Wild Ox)

1. *Midrash Tehillim*, ed. William Braude (New Haven: Yale University Press, 1959).
2. Yehuda Feliks, *The Animal World of the Bible* (Tel Aviv: Sinai, 1962).

Sparrow

1. *Midrash Tehillim*, ed. William Braude (New Haven: Yale University Press, 1959), p. 65.

Stork

1. Maurice H. Farbridge, *Biblical and Semitic Symbolism* (London: Kegan Paul, 1923), p. 81.
2. John Connolly, *Jerusalem Post*, 25 January 1988, p. 6.

Weasel

1. Henry Malter, *Treatise Ta'anit* (Philadelphia: Jewish Publication Society of America, 1928), p. 53.

Worm

1. *Midrash Tehillim*, ed. William Braude (New Haven: Yale University Press, 1959), pp. 315–316.
2. Bachya ben Joseph, *Duties of the Heart*, vol. 2, trans. Moses Hyamson (Jerusalem: Boys Town, 1965), p. 157.
3. Avigdor Miller, *Rejoice O Youth* (New York: Miller, 1962), p. 89.

Bibliography

Abrahams, Israel. *The Book of Delight*. Philadelphia: Jewish Publication Society, 1912.

——. *Companion to the Daily Prayer Book*. London: Eyre & Spottiswoode, 1932.

Albo, Joseph. *Ikkarim*. Ed. Isaac Husik. Philadelphia: Jewish Publication Society, 1929.

Asher, Asher. *The Jewish Rite of Circumcision*. London: Shapira Valentine, 1873.

Bachya ben Joseph. *Duties of the Heart*. 2 vols. Trans. Moses Hyamson. Jerusalem: Boys Town, 1965.

Braude, William, ed. *Midrash Tehillim*. New Haven: Yale University Press, 1959.

Brown, F., Driver, S. R., and Briggs, C. A., eds. *Hebrew and English Lexicon* (B.D.B.). Oxford: Clarendon, 1929.

The Disputation. Manchester, England: Scholarly Publications, 1972.

Driver, Samuel. *International Critical Commentary on the Holy Scriptures*. Edinburgh: Clark, 1895.

Epstein, Isidore, ed. *The Talmud*. London: Soncino, 1935.

Farbridge, Maurice H. *Biblical and Semitic Symbolism*. London: Kegan Paul, 1923.

Feliks, Yehuda. *The Animal World of the Bible*. Tel Aviv: Sinai, 1962.

Ferguson, Walter. *Living Animals of the Bible*. New York: Scribner's, n.d.

Fisher, Jonathan. *Scripture Animals*. New York: Weathervane, 1977.

Garstang, J. *The Fundamentals of Bible History*. London: Constable, 1931.

Gaster, M. *Ma'aseh Book (Sefer Hamasiot)*. Philadelphia: Jewish Publication Society of America, 1934.

Ginzberg, Louis. *The Legends of the Jews*. Philadelphia: Jewish Publication Society, 1946.

Goldman, Solomon. *The Book of Human Destiny*. New York: Harper & Brothers, 1949.

Hertz, J. H. *The Pentateuch*. London: Soncino, 1960.

Hirsch, Samson R. *The Pentateuch*. New York: Judaica Press, 1971.

Hirschfeld, Hartwig. *Al Khazari*. London: Cailingold, 1931.

Jastrow, Marcus. *Dictionary of Targum, Talmud and Midrash*. New York: Pardes, 1950.

Jellinek, Adolf. *Beth Hamidrash*. Vienna, 1878.

Jung, Leo. *Guardians of Our Heritage*. New York: Bloch, 1958.

Kleinman, Moshe. *Or Yesharim*. Lodz: Mesorah, 1934.

Lauterbach, Jacob. *Mekilta De Rabbi Ishmael*. Philadelphia: Jewish Publication Society, 1949.

Lewysohn, Ludwig. *Die Zoologie des Talmuds*. Frankfurt: J. Baer, 1858.

Maharal. *Kol Kitvei Hamaharal*. Ed. Abraham Kariv. Jerusalem: Mosad Harav Kook, 1948.

Maimonides. *The Guide for the Perplexed*. Ed. M. Friedlander. London: Routledge, 1928.

Malter, Henry. *Treatise Ta'anit*. Philadelphia: Jewish Publication Society of America, 1928.

Marmorstein, Arthur. *Studies in Jewish Theology*. London: Oxford University Press, 1950.

Miller, Avigdor. *Rejoice O Youth*. New York: Miller, 1962.

——. *Sing You Righteous*. New York: Miller, 1971.

Montefiore, C., and Loewe, H. *A Rabbinic Anthology*. Philadelphia: Jewish Publication Society, 1960.

Newman, J. *Agricultural Life of the Jews in Babylonia*. London: Oxford University Press, 1932.

Newman, Louis I. *The Hasidic Anthology*. New York: Bloch, 1944; Northvale, NJ: Jason Aronson, 1987.

Rosenbloom, Noah. *Studies in Torah—Luzzatto's Ethico-Psychological Interpretation of Judaism*. New York: Yeshiva University Press, 1965.

Seale, Morris. *The Desert Bible*. London: Wiedenfeld & Nicholson, 1974.

Schechter, Solomon. *Studies in Judaism, 2d. series*. Philadelphia: Jewish Publication Society, 1908.

Shoshan, Arieh. *Animals in Jewish Literature*. Rehovot: Shoshanim, 1971.

Singer, S., ed. *Authorised Daily Prayer Book*. London: Eyre & Spottiswoode, 1950.

Smith, W. Robertson. *Kinship and Marriage in Early Arabia*. London: A. C. Black, 1903.

——. *Lectures on the Religion of the Semites*. 1889, 1894.

Thieberger, Solomon. *King Solomon*. Oxford: East and West, 1947.

Toperoff, Shlomo. *Lev Avot on Pirkei Avot*. Jerusalem: Toperoff, 1984.

Trachtenberg, Joshua. *The Devil and the Jews*. Cleveland: Meridian, 1961.

Tristram, H. *Natural History of the Bible*. London: Society for Promotion of Knowledge, 1898.

Index

About the Author

Shlomo P. Toperoff was born in London in 1907 and after studying for several years at Eitz Hayim Yeshiva, won a scholarship to Jews' College and graduated from University College, London, with a B.A. in Semitics. He was later ordained as a rabbi by the London Beth Din. He served as a minister to the Sunderland community and then as regional rabbi to Newcastle Upon Tyne until his retirement. During these years he was very active in Jewish education and was well-known as a broadcaster, lecturer, and an exponent of the Jewish cause among non-Jews. Rabbi Toperoff is the author of the following books: *Eternal Life,* a handbook for the mourner; *Ehad Mi Yodea,* questions and answers on Jewish life; *Lev Avot,* studies on *Ethics of the Fathers;* and *Bishop Henson and the Jewish Problem.* He is currently enjoying his retirement in Israel with his wife, three children, ten grandchildren, and twenty-four great-grandchildren.